AWKWARD INTELLIGENCE

# AWKWARD INTELLIGENCE

## WHERE AI GOES WRONG, WHY IT MATTERS, AND WHAT WE CAN DO ABOUT IT

KATHARINA A. ZWEIG
TRANSLATED BY NOAH HARLEY

The MIT Press
Cambridge, Massachusetts
London, England

Original title: *Ein Algorithmus hat kein Taktgefühl. Wo künstliche Intelligenz sich irrt, warum uns das betrifft und was wir dagegen tun können*, by Katharina Zweig

© 2019 by Wilhelm Heyne Verlag, a division of Verlagsgruppe Random House GmbH, München, Germany

The translation of this work was supported by a grant from the Goethe-Institut.

This book was set in Adobe Garamond Pro by Jen Jackowitz. Printed and bound in the United States of America.

Library of Congress Cataloging-in-Publication Data

Names: Zweig, Katharina A., author. | Harley, Noah, translator.
Title: Awkward intelligence : where AI goes wrong, why it matters, and what we can do about it / Katharina A. Zweig ; translated by Noah Harley.
Description: Cambridge, Massachusetts : The MIT Press, [2022] | Includes bibliographical references and index.
Identifiers: LCCN 2021060550 | ISBN 9780262047463 | ISBN 9780262371858 (pdf) | ISBN 9780262371865 (epub)
Subjects: LCSH: Artificial intelligence. | Artificial intelligence—Philosophy.
Classification: LCC Q334.7 .Z84 2022 | DDC 006.301—dc23/eng20220521
LC record available at https://lccn.loc.gov/2021060550

10   9   8   7   6   5   4   3   2   1

To my mother, who helped me become a scientist, a teacher, and a writer.

# CONTENTS

The most important thing about this book, dear reader, is you! That's because artificial intelligence, AI for short, will soon find its way into every corner of our lives and make decisions about, with, and for us. And for AI to make those decisions as well as possible, we all have to think about what actually goes into a good decision—and whether computers can make them in our stead. In what follows I take you on a backstage tour so that you can see for yourself just how many levers computer and data scientists are actually pulling to wrest decisions from data. And that's where you come in: what matters at moments like these is how *you* would decide. That's because society should leave its important decisions to machines only if it is confident those machines will behave according to its cultural and moral standards. This is why more than anything else, I want this book to empower you. I hope to dispel the sense of helplessness that creeps in when the conversation turns to algorithms; to explain the necessary terms and point out how and where you can intervene; and finally, to rouse you to action so that you can join computer scientists, politicians, and employers in debating where artificial intelligence makes sense—and where it doesn't.

And how is it that artificial intelligence will soon find its way into every corner of our lives, you ask? For one, because AI can make things more efficient by relieving us of the burdensome, endlessly repetitive parts of our work. Yet I also see a tendency at present toward thinking AI should make decisions about people. That might occur when using data to determine whether a job applicant should receive an interview or a person is fit enough for a medical study, for example, or if someone else may be predisposed to acts of terrorism.

How did we get here in the first place, to the point where it became possible for so many of us to entertain the notion that machines are better judges of people than we ourselves are? Well, for starters computers are clearly capable of processing data in quantities that humans cannot. What strikes me, however, is a present lack of faith in the human capacity to judge. It's not as though we first came to perceive humanity on the whole to be irrational, liable to manipulation, subjective, and prejudiced when Daniel Kahneman was awarded the Nobel Prize in 2002 for his research on human irrationality, or more recently with the introduction of Richard Thaler's concept of nudging in 2017.[1] In our dim view of human judgment, it is, of course, always other people who are the irrational ones— all the more so if they have utterly failed to appreciate us for the highly individual and complex beings we are![2] This in turn leads us to hope that machines will unerringly arrive at more objective decisions and then, with a bit of "magic," will discover patterns and rules in human behavior that have escaped the experts thus far, resulting in sounder predictions.

Where do such hopes spring from? In recent years, teams of developers have demonstrated that by using artificial intelligence, computers are able to solve tasks quickly and effectively that just two decades ago would have posed a real challenge. Every day, machines manage to sift through billions of websites to offer up the best results for our search queries or to detect partially concealed bicyclists and pedestrians in images and reliably predict their next movements; they've even beaten the reigning champions in chess and go. From here, doesn't it seem obvious that they could also support decision-makers in reaching fair judgements about people? Or that machines should simply make those judgements themselves?

Many expect this will make decisions more objective—something that is also sorely lacking on many counts. Take the United States, one country where algorithmic decision systems are already used in the lead-up to important human decisions. In a land that holds 20 percent of the official prison population worldwide, and where African Americans are roughly six times as likely to be imprisoned as white people, one could only wish for systems that would avoid any and all forms of latent racism—if possible, without having to raise spending significantly. This has led to the use of risk-assessment systems, which estimate the risk that someone with a previous conviction runs of becoming a repeat offender. The algorithms work by automatically analyzing properties that are common among known criminals who go on to commit another offense, and rare among those who don't. I found it deeply unsettling when my research was able to show that one commonly used algorithm in the US resulted in mistaken judgements up to 80 percent of the time (!) in the case of serious crimes. Concretely, this means that a mere one out of every four people the algorithm labeled as "high-risk repeat offenders" went on to commit another serious offense. Simple guesswork based on the general likelihood of recidivism would only have been slightly less accurate, and at least had the advantage of consciously being pure conjecture.

So what's going awry when machines judge people? As a scientist coming from a highly interdisciplinary background, I consider the effects and side effects of software from a particular angle: socioinformatics. A recent offshoot of computer science, as a discipline socioinformatics draws on methods and approaches from within psychology, sociology, economics, statistical physics, and (of course) computer science. The key argument is that interactions between users and software can only be understood when seen as part of a larger whole called a *sociotechnical system*.

For over fifteen years now, my research has focused concretely on how and when we can use computers—and, more specifically, exploit data, or perform data mining—to better understand the complex world we inhabit. That lands me among the ranks of those with the sexiest jobs on planet Earth, even if a weekend spent wading through endless streams of data, sifting for exciting correlations with statistics, may not exactly sound like your idea of fun.[3] Personally, I can't imagine anything better! Yet at the start of my career, I used statistics without really understanding it, always uncertain of whether this, that, or the other method could actually be applied to data to yield interpretable results. This was due to the fact that after graduating high school I initially chose to study biochemistry, a course of study that typically spends little time on mathematics. We learned the basics of biology, medicine, physics, and chemistry—but not a single hour of statistics. They were probably hoping it would seep into our brains by pure osmosis if only we cooked up enough of the lab experiments they assigned.

Later, I came to bioinformatics, an entirely new course at the time that taught us to design and apply methods for examining the biodata that was then piling up in droves. Yet here, too, statistics was missing. Nor for that matter did either course provide any instruction in scientific theory, a baffling and dangerous blind spot present in the curricula of nearly every discipline in the natural sciences that aims to produce facts.

Under such circumstances, it should come as little surprise that many computer scientists and engineers are all too sure about their methods obtaining the pure and unadulterated truth from data. Especially with data mining and machine learning (the basis for artificial intelligence), they purport to have discovered the magic formula for solving each and every complex problem. For someone unaware of the fact that she is simply busying herself with models and can never achieve certainty once and for all, it is all too easy to rush into pronouncements like the following:

> Imagine a world where you can maximize the potential of every moment of your life. Such a life would be productive, efficient, and powerful. You will (in effect) have superpowers—and a lot more spare time. Well, such a world may seem a little boring to people who like to take uncalculated risks, but not to a profit-generating organization. Organizations spend millions of dollars managing risk. And if there is something out there that helps them manage their risk, optimize their operations, and maximize their profits, you should definitely learn about it. That is the world of predictive analytics.[4]

And that was just the introduction to *Predictive Analytics for Dummies*! Things take a more serious turn when companies advertise data mining software for "predicting hiring success" with phrases like this: "In the end the predictive possibilities are virtually unlimited, provided the availability of good data . . . let's take the emotion out of the hiring process and replace it with a data-driven approach!"[5] The catchphrase *data-driven* reminds me, however—there's someone here who would like to introduce himself and offer his services. His name is Artie, and he'd like to serve as your fully data-driven companion throughout this book. Artie is an artificial intelligence (AI) and may still be a bit slow on the draw when it comes to truly understanding humans. But he gets an A for effort!

As I'm sure you've guessed from the preceding two quotes, I would caution against the bold confidence of some—and against placing too much trust in Artie. Over the course of this book, I point out the situations in which we shouldn't breezily accept the results of machine learning without further ado. I then go on to make concrete suggestions about when algorithmic decision-making (ADM) systems are unacceptable, whether out of technological or societal considerations. At the same time, it's important to understand the enormous potential of *data mining*, by which I mean processing data with algorithms. In situations where ADM systems are permissible, I thus point out specifically where we have to be on guard, while also

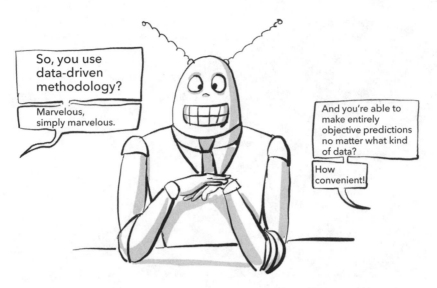

Meet Artie, an artificial intelligence system who would like to serve as your guide throughout this book. He can do quite a lot but he's still a little unsophisticated, so please be nice to him!

presenting ideas for how such systems can be developed, monitored, and regulated so that they really do make the best decisions possible.

This book gives you the information you need to understand how computers become judges over people, why they often fail to do this well when it comes down to it, and how we can improve them. Yet I also concentrate on situations where we don't want to use them in the first place, so as to avoid drawing conclusions about other people that may seem objective and like a sure thing but are ultimately wrong.

The book is made up of three sections. The first part introduces you to the scientific method for acquiring knowledge and sets you up with a tool kit for designing AI systems. In the second section, we go backstage to take a closer look at the ABCs of computer science—algorithms, big data, and computer intelligence—and how they relate to one another. Finally, the third section explores where exactly ethics enters the picture when it comes to computers and how we can best give shape to this process.

It's my hope that this book serves as a tool to help you get actively involved so that we as a society can make better decisions—with or without machines.

# THE TOOLKIT

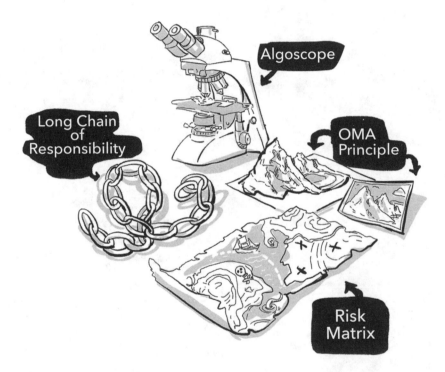

If you want to mess with artificial intelligence, you need the right tools. Going forward, the four tools described in this part will give you a method to work out the possible pitfalls if and when your boss or a state agency plans on using an algorithmic decision-making system—or, alternatively, to sound the all clear, because not everything that looks dangerous really is.

## ROBO-JUDGES . . . WITH POOR JUDGMENT

It wasn't the first time I had sat there dumbfounded by the results of our research, but it was probably the most memorable. My student Tobias Krafft and I had just finished sifting through the predictions made by a special software used in US courtrooms. We were horrified by just how bad they were—and yet the state was using them in a pivotal setting. The basic idea of using algorithms to predict whether a person will commit a crime or not calls the film *Minority Report* to mind. The movie is based on a short story from 1956 by the famous science fiction writer Philip K. Dick. In it, Tom Cruise plays a policeman who is able to identify and arrest potential criminals before they commit a crime, aided by "precogs," people with the gift of clairvoyance. What was a bizarre tale had now become a reality, albeit one where the predictive machinery was sadly lacking in precision.

Unlike in the film, real-world predictive software can of course neither "see" the actual crime nor know its exact timing. Instead, the software is fed basic information about the criminals it is meant to evaluate: how often a person has been arrested in the past, what kind of crimes they've committed, and information about their age and gender.

The computer than calculates a risk score based on the information, which you might compare to risk categories in car insurance, where higher-risk drivers are grouped into one category and lower-risk drivers into another. Yet a funny thing happens when a person is sorted into a category. Even if the driver hasn't done anything herself (yet), she receives the same treatment as others in her class. If those drivers were involved in multiple accidents, say, she pays more; if not, she pays less. This means that when a driver is first assigned to a risk category, what she pays isn't based on her own individual behavior going forward, but the past behavior of the people she resembles. In this way, financial risk is distributed among everyone within the same class.

How does that work when it comes to crimes that may be committed in the future? Well, the principle is the same to begin with. The computer identifies properties common among criminals who become repeat offenders and uncommon among those who regain their footing in society. It then uses those properties to determine a person's risk factor. In the case of car insurance, risk factors include the driver's age and the number of consecutive years without an accident. This isn't necessarily fair, and certainly lacks in complexity. Wouldn't it be better to conduct a personality test, and only after that decide the person's risk category?

It's argued, of course, that people are classified according to such highly schematic and easily measurable properties for efficiency's sake. At least with car insurance, however, the process is fair to the extent that any driver receiving his or her license at sixteen begins at the same starting point. Any subsequent classification depends exclusively on the individual's driving record and not that of their generation.

That wasn't something Tobias or I could say for the classification method used in COMPAS, the risk assessment system we were researching. Aside from information about previous crimes, an additional questionnaire asked prisoners whether their parents or siblings had committed crimes, or their parents had divorced early. While those circumstances may well leave a mark, they are hardly something for which a person is responsible or might alter themselves.[1] A criminal is thus evaluated and assigned a risk category based on whichever properties the software company deems relevant. If that person lands in a risk category where many of the people have committed another crime in the past, the software assumes that this person, too, will become a repeat offender.

The algorithm is advertised on the merit that it results in the right decisions around 70 percent of the time.[2] That number alone struck both Tobias and me as disturbingly low for software used by a public authority in court. In medicine, for example, such a low percentage would be considered unacceptable. Yet now we found ourselves face to face with results that proved how many people assigned to the highest risk category did actually relapse. The number was in fact somewhat higher than 70 percent for criminal acts in general—but only around 25 percent when it came to violent crimes. That meant that only one out of every four people who set off clear alarm signals as liable to commit another serious act of violence actually did. What was more, other colleagues had shown the average layperson to be capable of predictions that were just about as accurate.[3]

I've spent the last three years trying to understand why anyone would want to use algorithms that make such bad predictions and why governments would want to commission or purchase them. I also wanted to answer the million-dollar question, of course: How can we develop better software? And might there not be situations where algorithms shouldn't make decisions about people in the first place?

But what does any of this have to do with you, gentle reader? Isn't it all so technical that there's no room for any say in the matter? Your and my experience both over the past few years has been that we stand zero chance of changing the algorithms that help to define our lives. Google, Facebook, Amazon—it's all too confusing, too removed from the everyday. As individuals, but also at a societal level—certainly across Germany, maybe even Europe—we seem to go weak at the knees when confronted by the algorithms streaming across the Atlantic. The feeling of losing control owes in part to the fact that around the world, Google and company move in wherever they find the laws and regulations most congenial. Yet it's also due to the technology itself, which is often presented as an objective method for generating decisions—even Truth with a capital T—based on data. Such a perspective seems to leave us with one of two choices: Do we want algorithms that decide for us or not? Do we reject this whole digitalization thing, or do we sacrifice our personal data for all the new services it provides?

Fortunately, the algorithmic decision-making systems we will encounter in the decades to come are not the kind that will force us into this either-or scenario. Rather, we both can and should get involved in determining how these systems come to be applied by employers, educational institutions, insurance companies, and local government. In each of these instances, you have leverage. Whether as an employee, a student, a consumer, or a citizen, you can make your objections known and leave your own mark on how ADM develops. Granted, this only helps with the first of the two problems described with "Big Tech"; in this case, your partners are on site, so to speak, and accessible.

Are there really points at which you yourself can intervene at the level of content, though? To do so, it's essential to understand how the machinery behind artificial intelligence operates, especially *machine learning*. The process resembles something worthy of Doc from *Back to the Future*: it involves a great deal of tinkering around and is far less objective and self-guided than you might think. It results in decision-making machines that can be tweaked in any number of places and are held together by little more than string in many others. This also explains why it is so important to keep close watch over many of them—especially in the tinkering stage.

The potential uses of artificial intelligence will demand a wholesale political response in the realms of labor, education, and social justice. What this book asks specifically is how we want to design these machines; when we need to monitor them; where it's possible to do so; and, finally, the cases in which we're better off not using them in the first place. The truth is that

Figure 1
The process of machine decision-making is a result of professionally tinkering about with data and gadgetry.

as data scientists, we need you backstage with us, as employees, consumers, and citizens. This book assembles a toolkit that will get you on the job, and which I'd now like to introduce briefly.

THE TOOLS IN YOUR DECISION-MAKING KIT

The instruments described in detail over the following chapters will enable you to recognize three things: (a) whether you actually have to get involved; (b) if so, where you can intervene; and (c) the impact your perspective will have on the regulated use of machines. It isn't always necessary to get involved. To help you decide, I present the first tool in your kit, the *algoscope*, which helps us filter out which systems should be our primary focus of concern.

Are all systems that use AI suspicious per se? A great deal of thought has been given to questions like this in recent years. In their 2013 book *Big Data: A Revolution that will Transform how We Live, Work and Think*, Viktor Mayer-Schönberger and Kenneth Cukier propose a sort of overarching algorithmic safety administration that considers each and every algorithm coming to market.[4] As I show later on, this is neither sensible nor necessary in that particular form for a number of reasons. Mainly, however, it isn't necessary because not every ADM system needs to be brought before the witness stand. By and large, *the only systems* that call for regulation and for their internal mechanisms to be monitored are those making decisions about the following:

- People
- Resources that concern people
- Issues that affect people's ability to participate in society[5]

A small portion of all possible algorithms, in other words. The algoscope lets us focus on the ADM systems that carry ethical implications. Parts II and III of this book explain in detail why it is essentially only these systems that require tighter control and regulation.

What does this look like concretely? Systems that decide whether or not a screw is defective and should be taken off the production line do not fall into this category, nor for that matter does a system that distributes fertilizer over a field with pinpoint accuracy. A self-driving car that could potentially get into an accident, on the other hand, definitely makes the

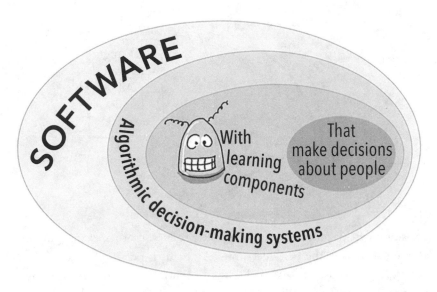

Figure 2
The *algoscope* helps us describe which kind of software we have to keep a closer eye on: algorithmic decision-making systems that directly or indirectly affect humans.

list. Systems that only recognize images or translate languages tend not to belong—unless of course they are built into self-driving cars, where they may lead to an accident. AI systems in the realm of medicine definitely belong, although it is less systems that recommend over-the-counter products than those that make decisions about treatment.

When your AI alarm bell goes off, then, first consider what the system is supposed to be deciding. If it neither directly nor indirectly impacts human well-being, you can return to the breakroom.

In cases where people's well-being is involved, the quality of a machine's decisions will depend on the following factors:

- The quality and quantity of data the machine has been fed
- The underlying assumptions about the nature of the issue at hand
- What society considers a "good" decision to be in the first place

A computer scientist might talk about this last point in terms of a *model for a good decision*; a philosopher would call it a kind of *morality*—that is, a set of standards or principles that any "good" decision should obey. For an algorithm to adhere to such a morality, however, the extent to which a given

decision does so must be made measurable to it; only then can a computer attempt to make "the best" decisions. And this is no simple matter. Suppose software is used to assign children to schools so as to make their way to school as short as possible. Does that mean the trek should be short on average? Or instead that a specific maximum length should be set for every child? Deciding how to later assess the quality of an algorithmic decision allows us to measure how good of a solution it really is. In computer science, this process is called *operationalization*.

To continue with the example of school assignment, there's also the question of precisely what kind of information the computer is being fed to calculate the distance from school. Are ideal travel times or actual travel times being taken as a basis? Has the walk to the bus stop been figured in? These types of decisions create the *model of the problem* that the computer is supposed to solve.

For the results of data processing to observe the moral principles we've established ahead of time, operationalization (O), the model of the problem (M), and the algorithm (A) must all work in concert. Together, they make up the second tool in your toolkit: the *OMA principle*. Beginning in chapter 2, we'll look at a number of examples that show what exactly this principle involves and how to go about using it.

Yet the OMA principle isn't sufficient on its own to determine whether and when machines should play a part in human decision-making. To do that, it's also necessary to consider their role in the overall process.

Figure 3 illustrates the long and winding process of developing and implementing algorithmic decision-making systems. It's a process I call the *long chain of responsibility*, one I explain step by step over the course of this book.[6] Ultimately, its length is problematic because it lets responsibility for individual decisions rest with so many people that later on it becomes difficult to hold any one person accountable. From the outset, though, it's important to recognize that *there are only a few points* along the chain where some form of technical know-how is necessary. By contrast, every step along the way includes aspects on which you can chime in. The long chain of responsibility weaves in and out of the topics raised in this book like a common thread, and with it you now have a third tool in your kit that shows where in the process you have to look.

Just how carefully we have to monitor a machine depends by and large on how much damage the decisions it is calculating can cause and how well we can shield ourselves from that damage. To this end, I present you with

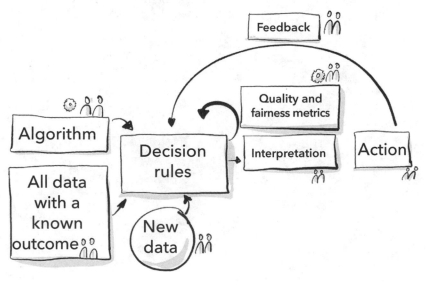

Figure 3
The long chain of responsibility. Only two links in the chain require some degree of technical knowledge, and you both can and should get involved at every step along the way. The following chapters discuss each individual step and what can go wrong in greater detail. A gear icon next to a box indicates some technical knowledge is necessary for the decisions involved at this step, while the icon with two people indicates that common sense is enough and/or social discourse is needed.

a fourth tool linked to a variety of control measures: a risk matrix that indicates how much regulation a given AI system might require. I'll explain this tool in greater depth with a couple of examples once we've completed the backstage tour.

With these four tools, your kit is complete. Once you've become a bit better acquainted with them, you'll be able to determine for yourself when they're called for.

Before we go backstage with AI, though, I want to make a quick detour to the basement, through the laboratories of the natural sciences. Why? Because the goal of artificial intelligence is to reproduce cognitive ability: in particular, the ability to draw conclusions about the world by observing it, that is, making discoveries based on data. And that, of course, is the grand domain of the natural sciences, something they've been doing for centuries now—and with great success.

In one sense, computers do this in a manner very similar to people; in another, they differ radically. To better explain, I invite you to travel back with me to the first time I was involved in a scientific discovery.

# THE FACT FACTORIES OF THE NATURAL SCIENCES

It's hot in the lab. Someone peeking in through the doorway would find me in a white lab coat, hunched over a stack of petri dishes and muttering numbers to myself: "1001, 1002, 1003." I'm counting the small shiny dots that have popped up on a nutrient agar plate, which indicate a yeast cell has divided so many times now that its offspring have formed cell clusters visible to the naked eye. "1004, 1005, 1006 . . ." Who would have thought that corroborating facts for an important biological discovery in cancer research could be so monotonous? When I first began a degree in biochemistry, I had somehow pictured it all as much more exciting!

At the time, my research group and I were looking into whether or not yeast cells underwent a simplified version of *apoptosis*, a biochemical process that is important in understanding how cancer develops. Also known as *programmed cell death*, apoptosis allows cells that are no longer functioning properly to break apart autonomously and wrap themselves up in tidy little packets, so to speak, to then be disposed of by the body's defensive cells. Apoptosis plays a critical role in multicellular organisms by guarding against malignant growths like cancer, for example, as damaged cells die off instead of spreading uncontrolled.

This makes it all the more important to understand the process, and yeast cells feature a number of properties that make it that much easier to research.

But yeast are single-celled organisms, which means they can't develop cancer to begin with. Why, then, should they be able to undergo apoptosis? What could the advantages for a unicellular organism be? These were the kind of questions I was hoping to sort out in my dissertation.

Our line of reasoning went as follows: A cell that simply dies without neatly wrapping itself up splits apart in uncontrolled fashion, releasing enzymes and other substances into the body's environment that may harm adjacent cells. A first advantage of the "packing job" in apoptosis, then, lay in preventing enzymes from nibbling away at their neighbors. Now it was my job to test for a potential second advantage: whether the little waste packets of an apoptotic cell might not serve as nourishment for other yeast cells. Yeast cells normally live in the immediate vicinity of their offspring, which means cells undergoing apoptosis could be recycled for future generations, to feed their grandchildren and great-grandchildren, as it were. That in turn would explain why an early stage of apoptosis can be found even within unicellular organisms.

To test this sort of scientific hypothesis, I needed an experiment that would measure the cells' ability to survive under different conditions. I began by letting the yeast cells multiply until they started to die off, then evaporated the liquid with their remains to create a concentrate.[1] This concentrate of apoptotic cells was then fed to a new yeast culture (a yeast culture is just a liquid medium in which a small number of yeast cells are initially allowed to grow undisturbed). Meanwhile, a second new batch— labeled the control group—received none. The goal was to see whether the yeast cells that received the concentrate multiplied more effectively than those that didn't. After a set amount of time had passed, we took small samples from both cultures and spread them out on nutrient agar plates. After leaving these to grow for a while, small but visible cell clusters, also known as colonies, began to form; those were what I was counting.

Yet the comparison between the control group and the group fed the concentrate wasn't entirely fair; a culture that has received some form of nourishment obviously stands a better chance of replicating than one that hasn't. So we tested a third group, giving it a concentrate that was prepared and administered in exactly the same way as the first experimental group except that it was made of young cells—that is, it was a colony that was still growing and was guaranteed not to contain any apoptotic cells.

My job was to count which of the three groups yielded more offspring— those fed the apoptotic cells, those fed the young yeast cells, or those that weren't given anything. So I sat there and counted—and counted, and counted, and counted. I ended up with three different distributions. A *distribution* describes how a certain property appears within a group; a wealth distribution shows how many people in a given population are millionaires and how many live below the poverty line, for example. In my case, the three survival distributions indicated which tests from each of the three situations showed at least one hundred or at least three hundred colonies. The results were astonishing. While both fed cultures did better than the unfed culture, as expected, those fed the concentrate of dying colonies were in fact much healthier than those fed young colonies. Up to eight times as many of the cells fed the dying colonies survived relative to the control group, compared with only three times as many for the cells fed the concentrate of growing cells.

Now that we had the data in front of us, we had to decide whether we could conclude directly on the basis of this observed difference that the concentrate of apoptotic cells had helped. It would have been ideal, of course, if nothing at all had grown on the plates with cells that weren't fed, only a small number of colonies had grown on the plates fed the young cells, and a great deal had grown on the plates fed the concentrate of old cells. Unfortunately, things in life are rarely so clear-cut that you can tell their differences apart immediately and beyond all shadow of a doubt.

Over the course of the experiment, I had set up multiple plates for both cultures that were fed a concentrate. Most of the plates fed the young cells showed fewer than a thousand colonies, while most of the colonies fed the dying cells showed many more. A number of plates in both groups, however, showed around a thousand colonies. Their distributions overlapped,

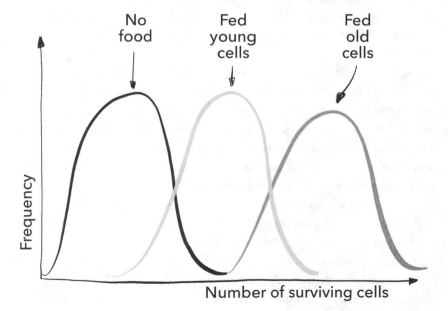

in other words; it was the *average* number of colonies fed the apoptotic cells that was significantly higher than the average for those fed young cells, roughly as shown in the illustration. Was the difference between the two averages large enough, though? Was it "statistically significant"?

The methods used to answer this type of question work in the opposite direction, by asking first whether it couldn't just as easily be coincidence that the plates fed apoptotic yeast cells did so much better. When seen from a purely statistical standpoint the number of viable cells in two different samples is bound to differ, after all, even if they are drawn from the exact same yeast culture. It's the same as if you roll a die a hundred times, then another hundred times. There is a large probability you will have rolled a different number of sixes in each round. Fortunately, statisticians know how large that sort of a difference normally is; when rolling dice, it is much more likely to be small rather than large.

It was the same with our yeast cells. If a statistician were to take a sample from two different yeast cultures, she would compare the *observed* difference in the number of viable cells to the difference that could be *expected* if the two samples were taken from the *same* culture. If the observed difference is comparable to the expected difference—that is, similar in size to what one could expect of two random samples taken from the same culture—the statistician would label that difference *insignificant*. In our case, the greater the variation between the two, the more strongly our results would support the hypothesis that one culture did in fact contain more viable cells than another.

Sadly, my degree in biochemistry didn't include a single course in scientific theory, nor for that matter any introduction on the right way to evaluate biochemical data statistically—and really, why bother?![2] Nor for that matter did all of us budding biochemists find mathematics exactly riveting. I myself had always liked math, but without any background in statistics I also lacked the knowledge necessary to demonstrate with any kind of scientific rigor that yeast cultures fed apoptotic cells did in fact show a statistically significant advantage over those fed younger cells.

I dove into the literature and soon found myself awash in a sea of statistics books. Yet nowhere did I find a patch of solid ground that would let me decide for certain which statistical method was the right one. How was I supposed to distinguish between methods for "normal distribution" and other types, for example? In the end, I opted for one of the simplest.

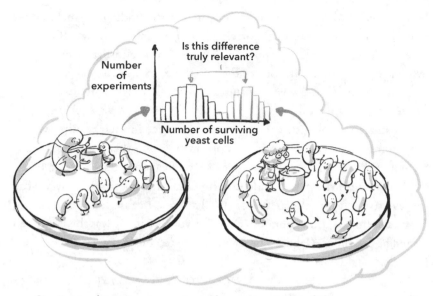

Number of experiments

Is this difference truly relevant?

Number of surviving yeast cells

## Statistical test measuring the meaning of the results (testing for statistical significance)

Figure 4
A test for statistical significance measures whether two observed distributions—in this case, the number of viable cells in two yeast cultures—show a conspicuous difference or not.

Armed with this knowledge, I became a one-eyed queen in the land of the blind. My team subsequently used the same method on any issue that didn't resolve itself quickly enough—always with our fingers crossed that we were actually doing it correctly.

As for my diploma thesis, it turned out that cells fed the concentrate of apoptotic cells did in fact stand a significantly greater chance of surviving. A "greater chance" is not the same as a guaranteed outcome, however, let alone sufficient evidence to conclude a direct causal relationship. It is merely a *correlation*—that is, two properties or patterns of behavior that are often observed to coincide with one another. What our observations did do was help support the hypothesis that a causal relationship *might* exist.

And so, after nine months of work, all I had managed to do was fit one small piece into a gigantic puzzle.

This explains in part why we've taken a quick detour through the laboratory on our way backstage. It's because the algorithms I discuss in this book would stop right here and simply accept the findings as such, rather than continuing on to test the correlations directly for causality. Rather, if algorithms find two things appearing alongside one another often enough it is made into a rule: "If you see the one thing, expect the other!" In this case the rule would be this: "Cultures fed by older cells will always show greater rates of survival."

Fortunately, biologists can shore up confidence in their results by running numerous similar experiments or drawing on other means of analysis and experimentation. That was exactly what my thesis advisor Frank Madeo did with the many students who came after me, and today we can rest assured that unicellular yeast cells "have good cause for apoptosis," as Frank and his coauthors phrased it.[3] As for yours truly, it was the last time I would be caught in a laboratory. I was drawn to computer science instead.

### FROM DATA PRODUCER TO DATA ANALYST

To this day, the joy I find in searching for the best ways to evaluate data has never left. Yet neither has the question of when and where a given method can in fact be used to meaningfully interpret results. This sort of critical awareness of methodology is called *literacy*, a term that encompasses a great deal besides: knowing the facts, but also selecting them discerningly with an eye toward solving a problem, as well as the ability to solve problems in and of itself.[4] These also happen to be the very skills needed in the field of artificial intelligence, where sometimes it is about as clear as mud which particular method will bring forth the best conclusions from the data.

As my work in biochemistry drew to a close, it was my time spent toiling in the laboratory I was happiest to leave behind. Generating data was *laborious* in every sense of the word, while the part that actually brought me joy—analyzing data—always seemed to get the short end of the stick. I found it maddening just how many individual experiments and observations it took to piece together a single causal chain. By *causal chain*, I mean setting facts in a sequence that explains how a given observation came to be. And that is exactly what machine learning promises us today: that *correlating* data with observed behavior is on its own enough to make decisions about new data.

To do so, however, would be to jump to conclusions. Tyler Vigen captured this in a particularly memorable way on his website, Spurious

Correlations (and in the eponymous book).[5] A visitor to the website encounters various sets of public government data, from which she then picks a set of her choosing—the number of divorces in Alabama, say—to see how the numbers have shifted over the years. Once a set of data has been selected, all other available data sets are sorted by their correlation with the chosen set. If two sets behave in the same way—that is, both sets of values rise and fall at the same time—they are said to be *highly correlated*. The exact degree of correlation can be measured using mathematical formulas. And lo and behold, the share of women with a degree in engineering shows a strong correlation with the divorce rate in Alabama![6] Figure 5 shows the change in divorce rate over time compared to the percentage of female engineers. They rise and fall nearly at the same time; in this case, there is a visible correlation. Was this undeniable evidence that women working in male professions destroys marriages? Or alternatively that women who had been left by their husbands were going on to pursue engineering degrees?

The answer in both cases is a resounding *no*. What is going on here is a case of *spurious* correlation, or coincidence that appears statistically.

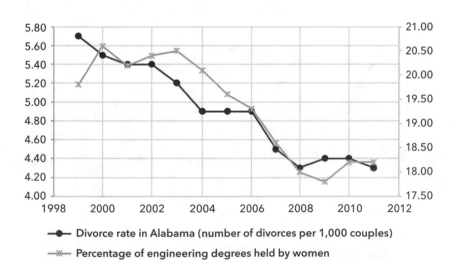

**Figure 5**
The annual change in the divorce rate in Alabama and the share of engineers who are female. The two curves show a strong correlation, which means they rise and fall nearly at the same time with only slight deviations.

Without first checking for potential causal links, then, the correlation in the figure doesn't allow one to make predictions about the divorce rate in Alabama based on the percentage of women graduating with engineering degrees (or vice versa!). Besides, there was another data set on Vigen's website that was even more closely correlated to the divorce rate in Alabama, and that was the number of lawyers (see figure 6).[7]

*Well, that could be cause and effect now, couldn't it?* I can hear you saying to yourself. *More lawyers offering their services and* presto, *now everyone gets a divorce!* The only issue is that the data was for the number of lawyers in the Northern Mariana Islands, an outlying US territory in the Pacific Ocean more than a day's flight from Alabama. Not exactly a prime suspect for a causal chain, in other words.

Nor is there any escaping this at a theoretical level: on its own, *a hypothesis that hasn't been tested does not count as fact.* Scientists only begin to talk about a *fact* in as many words after multiple hypotheses—each of which has been reviewed and could *not* be refuted experimentally—have been pieced together into a theory, which itself has led repeatedly to predictions that prove correct whether in controlled, replicable experiments or out in

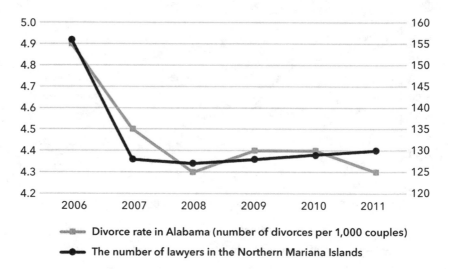

Figure 6
The change in Alabama's divorce rate over time shown alongside the number of lawyers—in the Northern Mariana Islands. The two curves reveal a stronger correlation than in the previous diagram; they cling to each other even more tightly.

the wild, so to speak. That's the scientific method (see figure 7)—the same method that those using algorithms with machine learning disregard when they apply algorithms' results directly as predicting future behavior. Later in the book, I return to discuss situations in which it is simply not enough to rely on correlations instead of establishing facts.

As far as my own journey through the sciences was concerned, however, I had seen the proverbial light at the end of the tunnel. Instead of gathering data, I wanted to better understand the methods used for its analysis, even develop those methods myself. By that time, I was already halfway through my course in bioinformatics and had fallen head over heels in love with theoretical computer science, which lays the groundwork for just that sort of analysis. I can still recall one of Professor Lange's first lectures in Computer Science III—Theoretical Foundations; he was describing the conversion of

## THE SCIENTIFIC METHOD

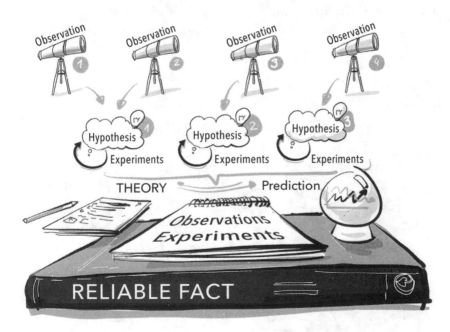

Figure 7
Any number of experiments must transpire before science begins to speak of a fact. Algorithms that use machine learning, on the other hand, leap directly from hypothesis to prediction, acting as though the hypothesis were itself a fact.

nondeterministic infinite automatons into deterministic finite automatons, or something equally abstract. I sat there, spellbound—though I couldn't say the same for many of my friends. Inspired by the great mathematician and computer scientist Alan Turing, theoretical computer science asks the philosophical questions raised by the discipline. What does computability actually mean? Do problems exist whose solutions only computers can compute or questions only humans can answer? Are there questions that neither people nor machines can solve by a general schema?

As it turns out, the first generation of computer scientists managed to come up with quite a surprising answer in response to these downright ethereal questions: *Based on everything we know to date, humans and computers are essentially capable of answering the exact same questions. Both are capable of solving—and fail to solve—the same problems.*

Such is the gist of the Church-Turing thesis.[8] Humans and machines can both calculate the root of one million or the shortest path from A to B, for example, or arrange a pile of books by the last and first names of their authors. Neither, by contrast, is capable of coming up with a general method for determining whether a given piece of software code will ever

enter an infinite loop. That's a shame, by the way, for a tremendous number of computer crashes could be avoided if such a method did exist. In this case, however, we've run up against the limits of computability.

Something about Professor Lange's lecture set me thinking. *People actually get paid to figure out philosophical and mathematical puzzles like these? That's what I want to do!* I found the work that went into designing algorithms especially appealing. Discovering, then analyzing and evaluating patterns in data was the piece of the puzzle that joined my various passions: my love of the natural sciences, but also my curiosity about what certain observations might herald for our lives and societies.

But does the Church-Turing thesis actually hold water? Don't we all share the sneaking suspicion that computers are much better at calculating than us humans, we who are constantly making mistakes? Only rarely do people get the same answer twice when asked to calculate even a small handful of numbers. We make subjective, not objective, decisions and often fail to see the forest for the trees. Fortunately, making calculations is child's play for a computer; adding together long strings of numbers, generating statistics, or searching for patterns within large sets of data—none of these pose a problem. Computers don't slip up; when given the same input, they will always give the same result. That's because the way a computer calculates the desired result has been prescribed by an algorithm that sets out in great detail how the computer should arrive there based on the input (more on that in chapter 3). No hormone fluctuations, no bad days, no surprise prejudices—they are *lifeless decisions*, in the best sense of the term.

Yet it is the very same spark of life that computers seem to be lacking when it comes to our deepest emotions and judgements as humans. Say you wanted to commission a poem or a piece of art: it's difficult to imagine a computer fabricating something that another person would enjoy. The same holds true for questions of justice and fairness in court, for example, or when educating our children or caring for the sick and elderly. Isn't a computer bound to fail in these instances if it "lacks soul"?

But these days, an entirely new breed of algorithm has emerged that would seem to overtake us in these matters as well. I'm talking about algorithms that make use of *machine learning* and that form the basis of artificial intelligence. With their help, texts that have foiled other methods for years are now suddenly translatable; the famed Babel fish from Douglas Adams's *The Hitchhiker's Guide to the Galaxy* seemingly draws nigh. Machine learning is capable of identifying the most important objects in a photo and transcribing spoken language more quickly and reliably than humans are able to. AI has even composed poems and painted pictures that humans regard aesthetically.

So why dally in the natural sciences when artificial intelligence seems like such a safe bet? It is because machine learning—an essential component of artificial intelligence—turns fact-finding completely on its head, a process that may have advanced very slowly over the centuries, but with great success. Instead of searching for reasons (a causal chain), machine learning identifies modes of behavior or properties that often appear alongside (correlate with) a significant event: the age of a driver in the case of an accident, for example, or personal characteristics often associated with criminal recidivism. Yet in contrast to classical algorithms, where a model is constructed (i.e., a mathematical problem is defined) *before* the algorithm is designed and used, it is now the algorithm itself that *constructs a model of the world from the data*; more on that later too.

The automatically discovered correlations I discuss in the following chapters are rarely reviewed and never examined for causal connections, yet they are still used to stick people in different risk categories. As we will see, this regularly leads to mistakes. To my mind, that means we can only take meaningful advantage of the efficiency gains machine learning offers if we examine these correlations for causal connections, as is normally done in the natural sciences.

With these initial considerations in mind, we are now ready to step backstage. As the first step in the long chain of responsibility, I pick up with the ABCs of computer science: algorithms, big data, and computer intelligence.

# THE ABCs OF COMPUTER SCIENCE

Flip through the pages of just about any newspaper these days and there's a good chance you will come across at least one article featuring the terms *algorithm*, *big data*, or *artificial intelligence*. One reads in the *Guardian* of "Franken-algorithms," while the *New York Times* warns us that people may wind up "wrongfully accused by an algorithm." We'll get to that, but before we do, what exactly is an algorithm? How is it related to digitalization, and what does it have to do with big data? Part II explores these questions through the ABCs of computer science, beginning with A, of course, which in this case stands for *algorithm*.

# ALGORITHMS: INSTRUCTIONS FOR COMPUTERS

The word *algorithm* may be bandied about more frequently today than in years past, but it remains an unfamiliar term to many; I've even heard top executives talk about "a-logorithms." Might the slipup express hopes that the term comes from the Ancient Greek *lógos*? That can't be right, however; in Greek, *a* is the prefix for *not*, and algorithms behave logically, not illogically!

No less than the world's most famous algorithmic scientist, Donald E. Knuth, reports in *The Art of Computer Programming* that early on linguists did in fact try to derive *algorithm* from ancient Greek—namely, from the terms *álgiros* (painful) and *arithmós* (number). As with *a-logorithm*, however,

## Where does the term ALGORITHM come from?

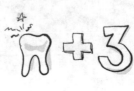

**A** The mathematician Al-Khwarizmi (9th century CE)

**B** From the ancient Greek *a* (not) and *lógos* (word, meaning)

**C** From the ancient Greek terms *álgiros* (painful) and *arithmós* (number)

Answer: (A) He wrote an important book about Indian numbers.

this derivation may well say more about linguists and their attitude toward something as scary as an algorithm than about the origin of the term itself. In truth, the word *algorithm* may be difficult to remember because it derives from the name of an Arab mathematician, Al-Khwarizmi, who wrote a foundational textbook on arithmetic in the ninth century. When it was translated into Latin three centuries later, his name was adapted without further ado to match the term's supposed roots in Latin or Ancient Greek. The confusion surrounding the word's origin explains the potential danger in mixing up *algorithm* with *logarithm* or *rhythm*, although neither of the latter is related, etymologically speaking.[1]

So what is an algorithm, exactly? You may be disappointed to learn that algorithms are simply fixed sets of instructions for solving clearly defined mathematical problems. The mathematical problem itself determines what information is available to the problem solver and the properties a result must have in order to count as a solution. In other words, the problem defines the relationship between *input* (the information entered) and *output* (the desired solution), which is why computer scientists are constantly referring to what is "given" and what is "searched for." Given the inputs of a street map and start and end points, for example, a navigation device

solves what's known as the shortest path problem using an algorithm that finds the shortest route from A to B.

An algorithm, then, is a detailed set of instructions for how to actually arrive at the desired solution once all the information needed to do so is available. It's essentially what you might explain to any rookie the first time she has to solve a common problem in her profession independently. In order truly to count as an algorithm, however, those instructions must be rigorous enough that they can be translated into programming languages. In computer science, this step is called *implementation*.

Math puzzles often bear a resemblance to the sort of mathematical problems I am talking about. I'll give you an example: Which four numbers add up to 45 and are also all the same number if you add 2 to the first, subtract 2 from the second, divide the third by 2, and multiply the fourth by 2?

In this case, the input is 45, and the solution—the four numbers—must meet certain requirements to count as solutions.[2]

And how would you go about solving the problem? Well, we can start by whittling down the list of candidates with a couple of initial considerations, and then proceed by trial and error. The difference between the first two numbers must be 4, as adding 2 to the first gives the same number as if you subtracted 2 from the second. It follows from this that both numbers are either even or odd. As for the third and fourth numbers, the fourth number equals one-half of the third number when doubled, which means it must be one-quarter as large as the third number. Any number multiplied by 4 comes out even, so for all four numbers to add up to an odd number (45), the fourth number must itself be odd. That means it could be 1, 3, 5, 7, and so on; once we've figured out which, the other numbers will follow. If the fourth number is 1, that makes the third number 4, and the first and second 0 and 4, respectively. Yet that would make for a sum total of 9. Proceeding by this kind of *guesswork*, we eventually arrive at 5 for the fourth number, making the third number 20, the first 8, and the second 12, for a grand total of 45.

Now, this kind of guesswork isn't an algorithm but what is called a *heuristic*, a term which comes from the Ancient Greek *heurískein*, "to find" or "to discover." Heuristics are strategies developed for finding solutions to a problem that have proven themselves so far, but offer no guarantee of ultimately finding a solution that satisfies all the set conditions.

One interesting example of a heuristic comes from ants in their own attempt to solve the shortest path problem. When off in search of food, an ant initially wanders about quite at random, leaving a scented trail behind it. Should it happen upon something delicious, the ant then uses the same trail to find its way back to the nest. On the way back, the ant lays out another scented trail—the more delicious the food and the more of it there is, the stronger the scent. This in turn allows the food to be found by other ants, who then further strengthen the trail. Since the scent is evenly distributed equally in all directions, it is more concentrated within a curve than directly at its edge. That's because all of the little scent molecules radiate out and meet somewhere inside the curve. It's similar to perfume; if you spray some on your neck, chest and wrist, the greatest concentration will be found somewhere between those sources. This means that other ants near the center of the curve will be somewhat more attracted than those at the edge, so that the loops initially traced by the first ant become shorter and shorter. Ultimately this leads to a relatively short path, albeit not necessarily the shortest.

We will come across the concept of a heuristic again in chapter 5 on computer intelligence, as most of the methods used in those cases aren't algorithms at all. How about that for some specialized knowledge to boost your score on trivia night! Still, it's an important point to remember: Only algorithms are certain to find the best solution; heuristics can't make the same promise.

Nor is it worth taking the trouble to develop an algorithm that applies generally anyway, so long as we are dealing with individual cases like the numbers puzzle given previously. Rather, it is regularly recurring mathematical problems *of a general nature* that deserve attention. Examples might include questions about the root of any given number or the product of any given set of numbers—but also consulting a database for all the purchases a customer made last year.

We've now assembled all the pieces we need to explain the term *algorithm*:

> An algorithm for a specific mathematical problem is any description of instructions sufficiently detailed and systematic such that if transferred into code correctly, its implementation will calculate the correct output for any correct input.

The computer scientist in me would like to add that the algorithm must calculate its solution in finite time, thank you very much. It seems to defeat

Figure 8
An *algorithm* has a plan for finding the solution and guarantees that it is actually a
solution. A *heuristic* is a method that attempts to find a solution.

the purpose, after all, if we have to wait until the end of the universe to get
the answer! But enough small talk. Would you like to see one for yourself, a
real live algorithm? Step right up then, folks, as I present to you: the sorting
algorithm!

### THE ALL-PRESENT SORTING PROBLEM

Some of my favorite childhood memories are the afternoons spent at my
father's side, helping him sort through his collection of advertising stamps.
From 1880 to 1940, these stamps were used similarly to today's trading cards, as a reminder to customers of whatever product they had just
brought home. The stamps had no postal value but were often used to dec-
orate letters or simply assembled in large albums; it was these albums that
my father collected.

Unfortunately, the stamps' previous owners had often glued them into the albums instead of simply laying them down on the pages. This meant that before we could get down to resorting them, we usually had to detach the stamps from the albums by rinsing them with soap suds in a little tub then drying them off with blotting paper. The sorting itself was more of an interactive process. In the case of the stamp shown in figure 9, the conversation might have gone something like this: "Papa, how do I sort this stamp? By the product, Sachsenglanz, or the company, W. Stephan? And if it's by the company, should I use the W, or the S for Stephan?"

Our sorting rules were constantly being refined, as each stamp looked different and I had no idea which keyword my father might use the next time he looked one up. So I sat there and sorted through hundreds upon hundreds of stamps, initially by the first letter, and then within that pile by the second letter, and so on. Finally, there was a small enough number of stamps left over that I could simply arrange them as you might a hand of cards and sort them in the correct order.

Figure 9
An advertising stamp from the early twentieth century.

Even if I didn't know it at the time, by using this "algorithm," I was actually solving for the sorting problem!

**The Sorting Problem**

Given as input: A number of things with a set of properties and sorting rules that establish for all pairs of things exactly which must be sorted first and which second.

   Sought as a solution: A sequence such that all directly adjacent things meet the sorting rules.

In the case of my father's advertising stamps, for example, the sorting rule went as follows: Go by the company name if one appears on the stamp; otherwise, use the product name. If the company name contains a family name, use that; otherwise, sort alphabetically, beginning with the first letter of the first word.

And how do you come up with a general solution for sorting? Any number of algorithms could do the job. Here I mention only two: the insertion sort and ascending sort. Anyone who likes playing cards will be familiar with the *insertion algorithm*: you begin with one card in your hand, then sort each following card you draw into those in your hand accordingly, whether by suit, from lowest to highest, and so on. And you're done. That's all there is to it!

The set of instructions for *ascending sort* is somewhat longer, but no more complicated. I'll stick with the example of playing cards, even if no real card shark would play her hand as follows:

- Lay all the cards down in a row on a table, then walk along the table, always moving from left to right.
- Whenever you notice two adjacent cards in the wrong order, swap their positions.
- The first time you don't have to swap anything, you're done—obviously!

Ultimately, this simple set of instructions will ensure all the cards are sorted. And now for the all-important generalization: in principle, both algorithms will also allow you to sort books according to their length or author name, or line up kids in a kindergarten class from smallest to tallest! Both algorithms can sort anything at all, so long as they are given sorting rules.

You've now been acquainted with two different sorting algorithms. There are literally dozens of such algorithms, each one leading to *exactly the same solution*, albeit in varying amounts of time. Some are well suited to numbers that exist within a relatively narrow range; others are equally efficient for all types of information, be it text or numbers.[3] If you find yourself with six minutes to spare, you can actually see and hear these different algorithms at work in "15 Sorting Algorithms in 6 Minutes," a video produced by Timo Bingmann that gives an audiovisual depiction of different algorithms sorting numbers.[4] They're all music to my ears, though the radix sort has the catchiest "tune." The final algorithm in the video is not an algorithm at all, by the way. Bogo sort creates a sequence at random, then checks whether it happens to be correct. Those may be clear instructions, and from a statistical perspective the algorithm will stop at some point—but not necessarily. In other words, it's conceivable that it can run forever, should chance dictate. While this makes bogo sort something other than an algorithm, it is probably still many computer scientists' favorite; everyone knows it.

What makes algorithms special, then, is their unbelievably wide range of application. The algorithms I've just discussed and their close relatives can sort *anything*: websites according to the relevance Google assigns them; products according to their popularity; photos according to their average color value (that, too, is just a number saved on the computer). Simply put, these algorithms can sort anything and everything that can be saved on a computer. And therein lies their true superpower: they allow us to *model* any number of problems from the real world by turning them into abstract problems.

*Modeling* is a term we will encounter frequently throughout this book, as the step where we are able to shape how algorithms actually work—and where problems arise when done poorly. Every time my father told me something about how to arrange the advertising stamps—what the *sorting rules* were, in other words—he made a *modeling decision*. When he told me, "First comes the name," he could just as easily have made a different decision. By defining the rules, he was constructing a model of the easiest way for him to look up the stamps later on. A different model wouldn't have changed the *sorting algorithm*, however, but only the outcome; the sorting rules aren't part of the algorithm itself but can be seen as part of the input.

I'll give another example. When my father first taught me to play cards, my ten-year-old self would always sort them by the same rules. That wasn't a smart move, at least not when you were playing against my dad! I soon learned that I constantly had to change how I sorted my cards, so that the sly old fox couldn't guess how many trumps I had in my hand. Practice makes perfect, it would seem. Yet changing how I sorted the cards didn't change how I picked them up; I still started out with a single card, then drew one after another, arranging them according to whatever system I was using that time around.

What we have in this case is an algorithm (in this case, me, the card player) that knows what the sorting criteria are and uses them to check where a given thing (in this case, a card) should be inserted. The actual instruction, "insert the card where it belongs," remains the same independently of whatever the concrete sorting criteria might be.

Sorting criteria reproduce what matters to the user, and as such form one part of what goes into *modeling a mathematical problem*. Modeling plays such a crucial role in artificial intelligence that it's worth clarifying with one more example. Without further ado, then, I present to you the most famous algorithm there is for finding the shortest path from point A to point B, one which will allow us to solve a whole host of other problems if modeled correctly. How *do* our navigation devices work, after all, those wily little operators that guide us from here to anywhere?

THE SHORTEST PATH

In his thirty years as a foreign correspondent for *Stern* magazine, there were three things my father always made sure to get when preparing for a trip. First, a book from the *Kauderwelsch* phrasebook series, which qualified him to speak the local language more or less at the level of a precocious three-year-old.[5] The second was maps and the third a pile of travel guides. He was meticulous in planning his adventures abroad, one part of which naturally included calculating travel times between the various towns and cities on his itinerary.

These days, when my mother asks me where exactly it is that I'm giving next week's interesting talk and how I plan to get there, I usually can't tell her. Nor do I need to, for that matter; I can look up the exact location of the event on the train ride, then rely on Google Maps to get me there the next day. "If you don't know where you're going, you might not get there," Yogi Berra warns us—but so far, all's well that ends well![6]

To my parents, taking such a casual approach to one's travels is unfathomable, even negligent. I myself am content trusting in the solution to one of the most famous mathematical riddles out there: the shortest path problem. Anyone who uses a navigation device expects it to calculate the shortest route from a given start point to an end point in moments. In order to do so, however, the concept of what the shortest path is must be made measurable, as it could in fact mean different things. It could mean the length of the route in miles traveled, for example—but that isn't always the deciding factor. One could also use expected travel time as a gauge, but it could just as easily be actual travel time, which is known only once the trip has been completed.

In computer science, the process of *making a concept measurable* that is not primarily mathematical by nature is called *operationalization*.[7] As with *modeling*, we will encounter the term *operationalization* repeatedly throughout

Figure 10
Artie can't quite get the measure of love. To do so, he needs devices that will help him make a social concept like love measurable. *Operationalization* is the process of deciding which of these instruments to use; that is, the method that will allow for a given social concept to be made quantifiable.

this book as it is what makes many *social* concepts accessible to artificial intelligence in the first place: the relevance of a message or friendship, say, or criminal tendencies, creditworthiness, or love.

For the moment, let's take the length of the route in miles traveled as our criterion. How would the algorithm function in that case? As with the sorting problem, countless algorithms and variations for solving the shortest path problem exist. The most famous of these took Edsger Dijkstra (pronounced "Dike-stra") just twenty minutes to develop in 1956, as he later recounted in an interview.[8]

To do so, Dijkstra imagined a network of streets as a literal network, with their intersections representing the network's "nodes." Each line connecting different nodes is associated with a distance; if you are looking to be highly precise, you could also use addresses along a street as the nodes instead of the intersections. In that case, the analogy is weak because a series of nodes strung up in a row, like pearls on a string, wouldn't really be considered a network anymore—but hey, sometimes that's how it is with analogies! Besides, it's important in order for the algorithm to function properly, as that very address may be our starting point.

These kinds of detailed decisions, which determine exactly what information the algorithm will receive as input, make up one part of modeling the problem. Together, they will subsequently play an important role in an algorithm's *interpretability*—that is, whether its results can be interpreted and, if so, how.

In the case of real-world street networks, it would also clearly be important for a navigation device to know whether a street is one-way or two-way. As it turns out, we can model this information as well by assigning the appropriate direction to the lines running between nodes. While most streets will be represented by two lines—one in each direction—one-way streets will only receive one line. In Hamburg, however, there is a street that can only be taken into the city in the morning and in the afternoon can only be taken out of the city: say hello to Sierich Street, the urban shape-shifter! Our modeling must also take special cases into account, including this fantastic creature, which is assigned one direction for inquiries in the morning and the opposite direction if the request comes in the afternoon.

Once we have developed our network of streets and identified a starting point, the Dijkstra algorithm can calculate the shortest route from our place of origin to every other address on the map.

We begin by adding our starting point to the list of already discovered places. Next, the algorithm identifies *every single point* on the map that can be reached directly from there. In this case, that means every usable street leading from the starting point. We now know at least *one* route leading from the starting point to each of these newly discovered points, and mark down the length and how we got there for each. That may not be the *shortest path* to each of these points, however. Let's take a closer look.

We start in the town of Otter Hill, which—surprise, surprise—is located on top of a hill. One street leads to Ottertown via long switchbacks over two miles, while a second street follows a steep ravine for three quarters of a mile to the village of Otter Valley. From Otter Valley, a second road leads to Ottertown in half a mile. The shortest path from Otter Hill to Ottertown thus goes via Otter Valley. Yet we couldn't have found this second route from our first round of discovery. Sometimes, then, additional iterations—always

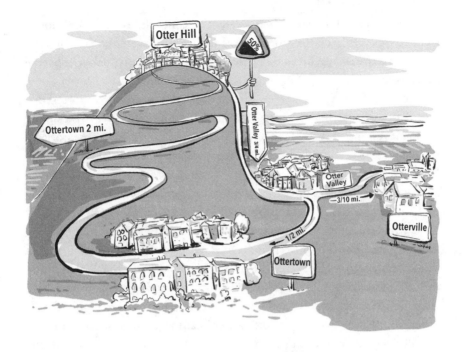

proceeding from points that have already been discovered—can reveal shorter paths to previously known locations.

By the same token, our initial round of discovery also made clear that there was at least one point for which we will never find a shorter path, no matter how many iterations we run. That is the point connected to our starting point by the shortest path, or street; all other streets connect to points that already lie further away. In our case, that point is Otter Valley. The distance between Otter Hill and Otter Valley is fixed, then, and we can continue on with our exploration from there. Some streets lead to points that have already been discovered; as we have shown, the street leading from Otter Valley to Ottertown shortens the overall route from Otter Hill to Ottertown. The road from Otter Valley to Otterville, by contrast, leads to a point the algorithm hasn't encountered yet. Otterville, too, is now added to the list of discovered points and the road there marked as the shortest route yet discovered. In subsequent rounds, this route could be replaced by a shorter one, as was the case with Otter Valley.

So in each round, out of all the points that have been discovered (and not yet used), we proceed with the point that lies the shortest distance from where we began. The rationale is always the same: there can be no shorter route to our starting point from any other place discovered so far. They are all at least as far away, and adding another road would only make the overall route longer.

And how do you go about identifying the point lying at the shortest distance from the starting point thus far that hasn't been used yet? You simply arrange every previously discovered point by distance, using an algorithm that solves the sorting problem. Sorting problem, you say? Piece of cake! It is in fact quite typical for one algorithm to make use of others; often the answer to one problem lies in solving multiple subproblems.

Thinking in terms of subproblems is one of computer science's great strengths. When computer scientists encounter a problem, we start by looking for the subproblems it contains. Once we've found efficient solutions to those problems, we then piece them together to give an overall solution to the original problem. As someone with a background in biochemistry, such an approach was wholly unfamiliar at first. Biochemists are highly detail-oriented and sensitive to context, which leads them to think in terms of overall systems. A biochemist would, for example, always keep the strong influence of the overall context in mind when determining which gene, and therefore which protein, is active in which cell under which circumstances. The resulting interaction between proteins can thus hardly ever be analyzed as an independent subproblem; it always depends on the status of the cell as a whole. Splitting apart a single problem into smaller problems then later piecing their solutions together demanded a mental revolution in the literal sense: a fundamental and lasting structural shift in how I thought over a short period of time.[9] It was fascinating to watch my brain adapt to an entirely new paradigm. At the same time, I've also become aware in retrospect that systematic thinking as I knew it from biochemistry is largely missing from computer science. It's something we could certainly do with in today's world as algorithms make a larger and larger impact on society.

Having sorted places on the map according to their (currently known) distances from our point of origin, we now have the final piece needed for Dijkstra's algorithm. For every step along the way, we select the point that currently lies the shortest distance from our starting point from among the list of known points, then search for points that are directly linked. This algorithm is used tens of thousands of times each year by beginning programmers. Like the sorting problem, the shortest path problem is of such a general nature that it serves as a model for any number of other issues. So what else can we do with it?

## COMPUTER SCIENCE AND CONSTRUCTING A MODEL

It's plain to see it's not just street networks that can be used as a basis; railway systems come to mind, for example. Be careful, though! Is it really that straightforward to adopt all the modeling decisions we've made for the street network just like that? Train stations would definitely be the nodes, but you probably wouldn't give the length of any given connection in terms of miles so much as the expected travel time. And how would you go about linking train stations to one another?

Let's give it a try. In the model, two train stations that at least one train travels to in direct succession are connected by a line that is associated with the shortest expected travel time. Take the cities of New Haven and New

Figure 11
Artie hasn't quite gotten the gist of the problem.

York, for example. If you travel with the regional Metro North line, it will take you about two and a half hours. That is much longer than the one and a half hours it takes to travel on the Amtrak Acela, so you enter the travel time for the Acela. If you do the same for every train station and every train, you wind up with a network that will allow you to calculate the length of the shortest train route from any given starting point to any destination. Sounds good, right? Watch out, though, when interpreting that output—that is, the length that's been computed based on the model. Our modeling isn't quite there yet.

Modeling the fastest train connection in the same way as a street network ignores several key factors: arrival and departure times, for one, but also possible layovers in between transfers. In practice, this will make the result completely useless, even if the algorithm itself is implemented correctly. The answer will give the route with the shortest travel time, but it will be based on the assumption that at every station along the way, one *could* in fact take the train with the absolute shortest travel time to the next station. Nor is the time spent waiting on the platform taken into consideration. The algorithm's recommendation for a connection between Philadelphia and Washington, DC, then, might look as follows: Take the Amtrak from Philadelphia to Baltimore at 10:00 a.m., followed by the Amtrak that leaves Baltimore for Washington at 7:50 a.m. Utter nonsense! My algorithm doesn't know that, however. Off she goes merrily calculating, unfailingly getting answers that are correct in and of themselves even with absurd modeling, but which make zero sense in the real world. Please keep this proviso in the back of your mind for later: modeling a problem always presents a problem in itself.

As far as the model for train timetables is concerned, there happens to be an elegant possibility that still allows the original algorithm to solve everything by comodeling which transfers are actually possible and the resulting waiting times. In the new model, I clone each train station for every time there is an arrival or departure. If it's possible to arrive at a station at 8:32 a.m. and leave it at 8:37 a.m., I create two nodes in the network and link them with a line that reads, "five-minute walk." I do this for every clone of that station, connecting each with a line whose length equals the difference between the two times. That may sound funny to us humans, but with these lines the algorithm now calculates travel connections that are actually interpretable. Figure 12 shows one section of a network created in this

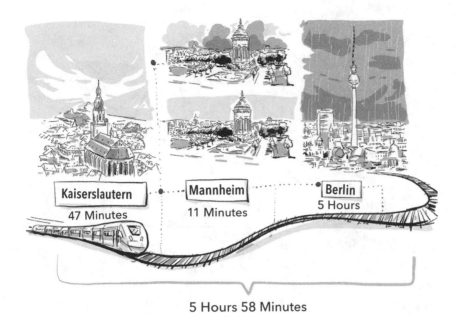

**Figure 12**
Only correct modeling will ensure that the results given by the shortest path algorithm can also be meaningfully interpreted. Here, the section shows that the first train from the city of Kaiserslautern takes forty-seven minutes to reach Mannheim, followed by a train that leaves Mannheim eleven minutes later and takes another five hours to reach Berlin.

way that generates feasible, real world travel connections when the Dijkstra algorithm is applied.

None other than the German railway system itself, Deutsche Bahn, notes that among others it uses a variation of the classical Dijkstra algorithm to solve a wide array of shortest path problems, including searches for the shortest connections based on number of kilometers traveled or expected arrival time, the cheapest options, the best connections, and the smallest number of transfers or changes of service.[10]

The sheer breadth of potential applications for the shortest path problem makes it a veritable jack-of-all-trades in computer science. Still, the truly important thing to realize here is that when this algorithm is given anything that resembles a street network in structure and a start and end point, it will begin calculating without a second thought. What an algorithm doesn't know, however, is whether the streets along the road map are at all

Figure 13
Artie comes up with an answer.

meaningful—that is, whether the lengths between nodes are actually inter-pretable. It's simply following orders. The subsequent interpretation of the results must always be carried out by a human, which means the model itself has to make sense. And who's supposed to check and see? Is that our job as computer scientists? To get a better sense of this, I'd now like to give you a quick overview of computer science as a discipline and its subfields.

WHAT COMPUTER SCIENTISTS DO WELL

Traditionally, computer science has been divided into three basic branches: technical foundations, theoretical foundations, and applied computer sci-ence. Each area demands a different kind of specialization and attracts a different kind of scientist. The technician is often a tinkerer, someone who is happiest making enormous machines jump at his or her command or cramming as much computing power as possible into a tight area. We have technicians to thank for everything from the fast chips in our smartphones to self-driving agricultural machines. Aside from studying computer science itself, those in the technical branch need a thorough grounding in physics and are always pragmatic engineers. They are without doubt the MacGyvers of the computer science world.

Practical, or applied, computer scientists deal with the question of how best to produce good software. To do so, teams develop models and devise methods they then use to create software that is as free of errors and efficient as possible. Other areas of applied computer science include designing databases that allow for large amounts of data to be efficiently stored and retrieved. They also develop what are called *distributed systems*, where a particularly involved calculation is divided up intelligently among multiple computers. To construct these systems in a meaningful way, one has to enjoy programming—and do a lot of it. The real pitfalls become recognizable only with a good deal of experience and require individuals who are able to take a bird's-eye view of the situation, think abstractly, and come up with simple solutions to complex situations. Applied computer scientists are the strategic yet pragmatic pathfinders, Princess Leia types bravely exploring what can be rocky terrain.

Finally, theoretical computer scientists, of which I count myself one, are without a doubt of the Professor Charlie Eppes variety, from the series *Numb3rs*. At times, the fundamental question driving this group—"What is the most efficient and quickest way to calculate?"—is just as philosophical as it sounds. Our research questions can be highly esoteric and without immediately discernable practical application. The theoreticians among us often reflect on systems that don't even exist yet—quantum computers, for example. A theoretician must take delight in the formal aspects of an issue, have a love of mathematics, and relish searching for patterns that algorithms can use. Theoretical computer scientists are responsible for designing algorithms and proving mathematically that they really do solve the problems they are intended to—and as quickly as possible, at that!

That part about mathematical proof turns out to be quite important, by the way. To understand why, let's go back to Edsger W. Dijkstra. Dijkstra was one of the first computer scientists ever to hold the title, once recalling in an interview that he hadn't been able to give his profession as "programmer" at his own wedding because the job description simply didn't exist yet![11] At the time, most programmers studied physics, electrical engineering, or mathematics. But I digress. In 1959, Dijkstra challenged a number of colleagues to a small competition, in which they had to solve a particularly thorny problem using algorithms. He received many answers, none of them right; in each case, he proved it to the author by finding a scenario in which their algorithm would go wrong. At some point, it all became too much work as the algorithms grew increasingly complicated and it took him longer and longer to find a case that brought the error clearly to light. So Dijkstra turned the tables: from now on, a submission would only be considered valid if a mathematical proof was furnished at the same time, showing that the algorithm was in fact correct.[12]

You might think this would make the challenge even more difficult. Yet Dijkstra's colleague Theodorus J. Dekker ended up solving the problem in just a few hours. The secret to his success sounds almost banal: It helps to start out by thinking through the properties a solution must actually have to prove it's correct. Once Dekker had recognized the formal requirements for the solution, the solution itself became easier to find.

And that is exactly what today's algorithm designers are trained to do: formalize a problem in such a way that it becomes absolutely precise, without any possible misunderstandings. We then try to find an algorithm that solves the problem by proving it does so always and for every possible input. Finally, we are interested in how long the algorithm takes to do so.

Theoretical computer scientists essentially start out by constructing a model of the world and the solution. I'll give an example. When my family pays a visit to my sister-in-law, we search for a route along the way that (among other things) includes rest stops with play areas for the kids. As a theoretical computer scientist, I could incorporate that into my model for the "shortest-path problem." The design for the mathematical problem, then, might include something like: "Given a road map with designated play areas . . ." Ultimately, the result is also just a model; in formulating the problem, I could attach a condition to the solution, such as: "The route must have a play area where we can stop every ninety minutes." In other words, my model for how people plan their travel itinerary describes how things in the real world are linked from my point of view, which details matter and which don't, and what property (or properties) a solution must have to count as a solution.

In theoretical computer science, this can lead to highly abstract models like the marriage problem: Given: two groups of people, where each individual has a list of all the people in the other group they could imagine marrying. Sought: A solution that allows for the greatest number of desirable marriages. From a mathematical perspective, this is a tremendously exciting problem with a range of fascinating variations—albeit one that, admittedly, *might* not do full justice to the world's complexity. But computer scientists are well aware of that. Well, most of us anyway.[13]

By making decisions like dividing marriage candidates into two groups, we define both the mathematical problem and the requirements for the solution ("as many marriages as possible"). We then check to see whether an algorithm exists for the mathematical problem that is able to calculate a solution for the data on hand—that is, a specific real-world situation. In a final step, the solution is interpreted in terms of the model and culminates in an action. Versions of the marriage problem are used on dating websites, for example, where an algorithm's answer results in the platform introducing users to each other.

Yet the connection between an algorithm's results and the ensuing action is not always so straightforward. The example of the train schedule above made it clear why the interpretation must be made to fit the underlying model. If we only consider the shortest possible transfer time between two stations, then the "shortest path" discovered by the resulting model will in all likelihood be entirely hypothetical. The action should not be to pass the results on to travelers. Similarly, our model of the marriage problem may not lead to the best solution available if it only allows people from different groups to marry; more pairs might have been formed had all the candidates' preferences been known. Overall, modeling must *always* provide the framework for results to be interpreted.

Despite this caveat, the overall series of steps needed to define and solve mathematical problems is relatively linear and easy to understand.

Is there an algorithm for every mathematical problem? No! As a matter of fact, there are mathematical problems that can be defined quite clearly, yet for which no general algorithm exists. I mentioned one such problem earlier—whether a given software might at some point enter an infinite loop. It would be truly marvelous if we did have a general algorithm for it, one that every phone app could simply run before going online. We would never have to watch the rainbow ball of death spin again! It's a shame, but

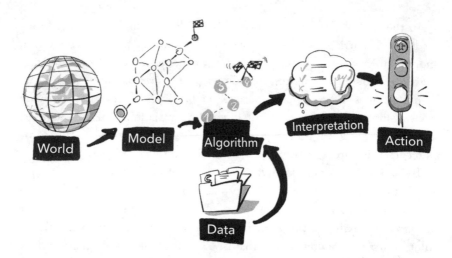

Figure 14
The classical method begins with a mathematical problem, which models one small
part of the world. An algorithm is developed for the problem, which is then fed data.
Finally, the solution has to be interpreted and usually results in an action.

nothing doing; it really doesn't exist. We can even prove it—though I'll
spare you the details here.

All in all, theoretical computer scientists get a kick out of mulling over
problems—we're especially well-trained for it. We know which methods
exist for proving that algorithms do what they're supposed to. But of course,
sometimes even the best laid plans go astray, as the next section shows.

CLASSICAL ALGORITHMS' SUSCEPTIBILITY TO FAILURE

A foreclosure is a devastating life event. The neighbors who just weeks ago
counted you among the ranks of respectable homeowners no longer greet
you. The kids have to switch schools and your family has to squeeze into a
rental. Colleagues at work ask what made you give up such a lovely home,
and your only reply is that the mortgage loan modification you so desper-
ately needed has been denied by the bank. Now you aren't able to pay back
your loan and your homeowner's credit falls through the basement. The
future looks bleak: you've lost your creditworthiness.

How would it feel, then, to discover years later that it had all come about
because of an error made by some algorithm used in automatic approval

software? Jose Aguilar knows.[14] In 2017, the Wells Fargo bank was forced to admit that over the course of seven years (!), between mid-March 2010 and the end of April 2017, 870 customers had been denied a normally unproblematic form of debt restructuring. The decision hinged on the question of whether the restructuring would lead to better terms for the customers, who would thus be more likely to actually pay the loan back. Under normal circumstances, the bank would agree to this kind of thing, but in this case the algorithm wrongly calculated the anticipated notary cost and automatically rejected the requests.[15] A total of 545 people lost their homes as result of the denials, among them Jose Aguilar, who had sought repeatedly to obtain a loan modification.

As you can see, these kinds of errors can have serious consequences. They can also be discovered relatively quickly—if one holds oneself to high standards. It's possible to implement classical decision-making algorithms with relative transparency if the programmer wants. By including user-friendly commentary in the code and making the right choices about how to structure the details, an algorithm's set of instructions can be designed to be easily understood. That happens less frequently than one might wish, but in principal classical algorithms' code can be crafted so that people can understand it. The main problem in the case of Wells Fargo was more likely a failure to notice that an entire series of homeowners had been wrongly judged in the first place. As grave as each individual story may be, 870 wrong decisions over seven years is a relatively low count when compared to the total number of cases the software probably evaluated. At any rate, as soon as false decisions are discovered, the reason for them is usually discovered quite quickly: it's what we computer scientists are trained to do.

## THE UNETHICAL USE OF ALGORITHMS

Aside from unintended side effects with potentially catastrophic consequences for society, classical algorithms can of course also be explicitly and purposefully programmed for unethical ends. That is clearest in the case of malware—password phishing or identity theft, for example, or taking entire hard drives hostage. Yet it would also be a piece of cake for me to build a navigation app that takes users past a large fast-food chain at least once a trip. It's just that there hasn't been enough demand to date for that particular gadget!

It doesn't end with imaginary examples, though. In 2018, it came out that a number of airlines had been using algorithms that comparatively often would assign separate seating to people who wished to travel together.[16] This let the company then charge a fee if the passengers wanted to sit together after all. Ryanair was singled out for engaging in the practice in an investigation by the UK's Civil Aviation Authority, although the company denies it.[17] In this context, it's worth considering that separating groups who are flying together isn't just a source of aggravation for the passengers but can actually lead to longer evacuations in cases of emergency. Such was the conclusion reached by the Royal Aeronautical Society Flight Operations Group, in a report that argued for always allowing families to sit together.[18] In this case, using the best software possible is thus an ethical requirement regardless of whether some people are not seated together intentionally or due to an oversight.

The best-known example of an algorithm being used unethically has to be the diesel emissions scandal at Volkswagen. In this case, the car company used software that recognized whether a vehicle was being used in a test situation or on the road. If the car was under inspection, various systems would switch either on or off to meet required emissions standards. Emission levels were quite a bit higher under normal operation, however, which improved the driving experience and saved on the costs a more complex emissions system would have entailed.[19] But who exactly is responsible for the errors an algorithm makes or its immoral behavior? How much of the responsibility falls specifically to the algorithm's designers?

## WHAT ARE ALGORITHM DESIGNERS RESPONSIBLE FOR?

How do we gauge the level of responsibility a developer holds for the results of the algorithms they design? For programming errors like the one that led to mortgage restructures being wrongly denied, one can hold the company to account even if it wasn't pursuing some dastardly scheme. On the other hand, developers can't do much in cases where an algorithm created for another company has been used incorrectly. If the proper way to use an algorithm's has been clearly communicated and users still enter the necessary information incorrectly, we are dealing with operator error. Ultimately, we have to check whether the algorithm itself was free from error and if a description for how to go about using it was available. The more a developer team knows about how exactly an algorithm will be used and the more say the team has about that usage, the greater the responsibility the team holds for the algorithm's subsequent behavior.

Yet it's also important to understand that even classical algorithms can be quite nontransparent in how they behave. By *nontransparent*, I mean that the precise behavior of the algorithm cannot necessarily be described in abstract terms. This was demonstrated quite effectively by Donald E. Knuth and Michael F. Plass in 1981, when the two sketched out an algorithm for improving the design of automatic line breaks in a body of text.[20] To do so, they developed a standard for the "beauty" of a text and an algorithm with a comparatively high number of parameters. Each combination of parameters resulted in slightly varied line breaks; the sheer number of possibilities made it impossible to say how the sentence would change if one were to change a single parameter. The authors wrote: "So many parameters are present, it is impossible for anyone actually to experiment with a large fraction of the possibilities. A user can vary the interword spacing and the penalties for inserted hyphens, explicit hyphens, adjacent flagged lines, [etc.] . . .] Thus one could perform computational experiments for years and not have a completely definitive idea about the behavior of this algorithm."[21]

Now at first glance, this seems to contradict the Church-Turing hypothesis. Aren't humans supposed to be able to calculate exactly what computers can? The answer is still yes. An algorithm consists of a series of very simple basic commands and checks that can just as easily be done by a person—albeit about a billion times more slowly. Knuth and Plass aren't arguing that humans couldn't determine how the program would respond to *individual* entries or what the outcome would be. Their point is that we can't summarize its general behavior abstractly. Will this or that parameter turn out a more aesthetically pleasing result? And how does that depend on the text itself? What Knuth and Plass are getting at, then, is that you can't describe an algorithm's behavior as you might describe a physical body in a three-dimensional landscape, where only a few rules are needed to predict general forms of behavior—the exact location a given body will come to rest when dropped, for example. With line breaks, the situation is more complicated as they will always depend on the following words. The last word of a paragraph, for example, may mean the first line must be broken up differently for the paragraph to retain an aesthetically pleasing form. These sorts of global interactions—the final word altering the position of another word somewhere else in the text—are an indication of a complex system whose behavior can be observed but will be difficult to predict using only a few abstract rules. Luckily, the aesthetics of line breaks hasn't sparked too much quarreling to date, although specialists exist even among the ranks of the nerds. Knuth and Plass's article speaks of nothing else for sixty-six pages while on Reddit, the typography forum has over 183,000 (!) members and counting. Outside of this community, however, a bad line break isn't considered a crime. But the phenomenon of it being impossible to describe an algorithm's behavior abstractly is one we will encounter again, especially when we turn to computer intelligence.

All this, of course, also makes it more difficult to discover errors in algorithms. Nevertheless, of the algorithms we have seen so far, it's still possible to check whether or not supposed answers do in fact satisfy the desired properties in the first place. Sometimes that's easy to do; sometimes it isn't.

An algorithm is said to malfunction when it calculates an answer that isn't actually a solution to the problem—a navigation app that doesn't always answer with the shortest route, for example, or seating assignments that divide groups more often than the seating plan itself requires. On the other hand, an algorithm that regulates a car engine differently than the law intends can't be said to malfunction if that was the problem the programmer intended to solve. In this case, it's *using* the algorithm itself that is wrong.

The important lesson here is that in principle, it becomes easier to detect these kinds of malfunctions first of all when the properties a solution must have to qualify as such have been precisely defined. So long as these definitions are made transparent—as with the navigation device that promises to show us the quickest route—in principle users can check the software themselves. Things become even easier if multiple implementations exist, all of which should arrive at the same solution as they are intended to solve the same mathematical problem. In this case, a simple comparison of the solutions can reveal any possible errors. "What? TomTom wants me to go left, but Google Maps says to the right?" Users can then check to see which algorithm is correct and actually has the shorter route.

**So who is responsible for how an algorithm is used? All in all, I would say this:**

The more information and control algorithm designers have regarding the precise context for an algorithm's use, the greater the responsibility they have for its results and their consequences. If my algorithm is used in only one context that is familiar to me, it will act in the way I have programmed it to. Accordingly, its results are also my responsibility.

## AT A GLANCE: THE OMA PRINCIPLE

In this chapter, I've shown that the results of a classical algorithm can only ever be understood for one defined context. That means we have to understand the interactions between the *operationalization* (putting social concepts in measurable terms), the general *modeling* of the context as a mathematical problem, and the *algorithm* itself. If the modeling isn't built to suit the algorithm, the algorithm will still come up with an answer, but it will be impossible to interpret it meaningfully. The example I gave here was modeling train connections without taking note of departure and arrival times. The algorithm won't recognize the modeling as false; it will still calculate the shortest route, but in most cases that won't result in a feasible travel connection. It's important to keep in mind that this particular modeling might well be totally reasonable for *a different question*. You might use simplified modeling if you wanted to know the *least* amount of time it takes to travel from one train station to another, for example.

A second important point in this chapter was that with every operationalization, we must ask whether other options existed for making the social concept measurable and whether this particular form of operationalization really captures the most important facets of the concept. In the case of the navigation device, this meant asking how we were actually going to define the shortest route—that is, the length of the route as opposed to the shortest expected travel time.

Finally, I discussed the distinct possibility that the algorithm itself is faulty. If an algorithm is given a mathematical problem whose solution has known properties, we can check to see whether or not possible answers actually qualify. We can spot a route that is presented as the shortest but isn't for example, just as we can quickly check to see whether an answer to a sorting problem does in fact follow all the sorting rules that have been defined. For most heuristics, that opportunity doesn't exist. Many of the methods in machine learning are heuristics, however, which is one of the reasons why it becomes difficult to verify whether the answers they find really are the best. Last but not least, we discussed the fact that algorithms

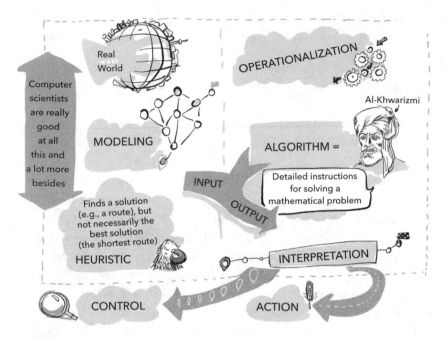

Figure 15
A summary of chapter 3.

can also be intentionally used in unethical ways. Figure 15 gives a visual summary of the chapter.

To understand what can go wrong when training an algorithm and how ethics can be introduced into the mainframe, so to speak, first I'll show you what is meant by the buzzword *big data*. And then—into the classroom with the machines!

# BIG DATA AND DATA MINING

If you want to get an idea of what people are searching for online at any given point, Google Trends is a fascinating tool that lets you view the search volume of specific terms.[1] The site doesn't show the absolute number of searches, but their relative share. To do so, Google first determines the largest number of search queries the term received within the period of interest, then gives the relative share throughout that period. If the greatest number of people searched for "algorithm" in January, for example—let's say one hundred thousand—then for every following month, the total number of searches for "algorithm" is divided by one hundred thousand.

Figure 16 illustrates the drastic shift in the relative number of searches for the basic terms of computer science over the past fifteen years. Whereas the search volume for the term "algorithm" tapered off relative to a high point in February 2004, searches for "big data" first really took off around 2011, a trend that reversed itself in early 2017. The search for "artificial intelligence" never really fell off, although there was far less interest in the term between 2005 and 2014 than there was in "algorithm" in early 2004. In the meantime, interest in the term "artificial intelligence" had quintupled as of 2019—nor have we likely seen the end. Artificial intelligence, it would seem, is back in style.

Google Trends is a tool that relies on *big data*, which means the underlying data was not collected in representative fashion but gathered from actual searches. People's interests are reflected in the terms they search for, although the two do not always necessarily go hand in hand.

The term *big data* is also meant to express the quantity of the underlying data, which is enormous. While only Google could likely give a definite answer, it's estimated that the company receives around 3.5 billion (!) search queries every single day.[2] The numbers circulating on the web vary greatly; some put the daily total as high as five billion.

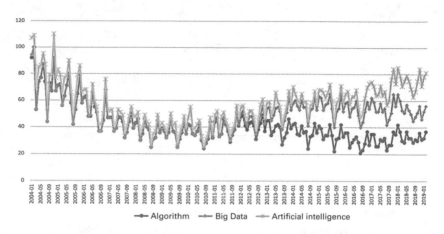

Figure 16
Relative number of Google searches for "algorithm," "big data," and "artificial intelligence."

For each term, Google Trends also states whether the user searched for something else that (among other things) either contained the search term or was closely linked to it. Such "related search queries" indicate that Google Trends results are noisy data. In computer science, *noisy data* means that not all the data present is relevant. As with a weak radio signal, one part is simply white noise, or static. In this case, not every search counted by Google Trends that included a term like "big data" or "artificial intelligence" also meant that the searcher was looking for information about those terms, strictly speaking. Instead, the search queries Google identified as "similar" suggest that some who searched for "AI" were looking for info on the film *A.I. Artificial Intelligence*. Others were searching for info about a book by Manuela Lenzen with the unsurprising title *Artificial Intelligence*. Of course, these more specific searches still indicate an interest in the subject matter, albeit one with a somewhat different orientation than the term itself.

> And there you have it, the three key characteristics of big data: it (1) deals with *large quantities* of information that (2) are often created and must be processed over *short periods of time* and that (3) usually comprise *different sources*.

This is summed up as *the three Vs*: *volume*, *velocity*, and *variety*. Others add two more Vs: big data is used when it is suspected that its analysis will bring *value*, but the data must be inspected for its *validity* in order to do so.

Whatever the case may be, big data can pile up anywhere there are sensors connected to the internet. And I'm not just talking about the sensors in the GPS system on your smartphone or the thermostat in your smart home, but the sensors installed in each and every camera, keyboard, computer mouse, and Siri- or Alexa-enabled device through which you actively or passively let information flow.

Big data can also amass over the course of any number of biology or physics experiments—with genome analysis, for example, or astrophysics. Recently, our first glimpse of a black hole made the rounds in an image that contained nearly five petabytes of information.[3] A petabyte—that's approximately one million billion bytes . . . !

Sorry, my head just exploded. I can't picture it. Can you? Even putting it in everyday terms doesn't make things much clearer—saying, for example, that a petabyte equals the total amount of data contained in every YouTube video uploaded on a typical day in 2014 (according to Google). Still inconceivable. Let's try another way. A modern computer built for home use has a storage capacity of one terabyte. It would take one thousand of these machines to store a petabyte. In the case of the black hole, the quantities of data required to store the image were so large that they couldn't reasonably be sent via the internet, and the memory drives had to be transported physically by airplane from observatory to data center.[4] In this case, what was accomplished is truly awe-inspiring.

Articles about how the image was constructed, moreover, quickly make clear just how excited the scientists involved were about their discovery. And suddenly Katie Bouman has become a household name—a newly minted professor of computer science whose algorithm made it possible to assemble the image in the first place. In one video where she describes what a team of nearly two hundred people was able to achieve, you can still glimpse the eureka moment scientists are always hankering after reflected in her smile and the twinkle in her eye. The video is absolutely worth a watch if you are looking to understand something about the thrill computer scientists feel upon discovering patterns in data.[5]

To my mind, science is hands down the best mix of detective game and earnest search for discovery. And I suspect anyone who pursues science based on large quantities of data would agree. It is truly exhilarating to be able to rummage about in data no one has seen in that form before. If you really want to make me happy, all you have to do is leave a fancy-schmancy pile of data outside my door, hand my family a week pass to Disneyland, and promise you'll keep me supplied with meals that can slide easily under the office door.[6] The thrill of poking about for fascinating patterns amid piles of data and the pure joy in discovering them can be difficult to put into words. But it is absolutely comparable to the rush others might feel on landing a business deal or winning in sports, or when everything in life simply falls into place. Circling in on one of these eureka moments is intoxicating; you try out every possible path, and if you stumble along the way you get up, dust yourself off, and try again. Sometimes that means late nights,

where enough sleep, a healthy diet, and social contact—and I'm serious—all become afterthoughts.

When I first began work on my dissertation in computer science in January 2003, it was quite difficult to come by this kind of data—impossible, in fact, for a theoretical computer scientist with no industry contacts. So when Netflix, a company that was little known in Germany at the time, staged a competition that gave access to over one hundred million ratings from more than 480,000 users for 17,770 films, it was like hitting the jackpot for those of us in the data science community.

### THE NETFLIX PRIZE: LIFE ON THE REALLY BIG SCREEN

Our mission, if we chose to accept it: develop an algorithm based on the data provided that would predict which ratings the same users had given other movies (which had been kept secret from us). It was as though we were being asked to design a crystal ball to predict that "Ms. Jones will simply love *Pretty Woman!*" or "Mr. Smith will like the movie *Titanic* just fine." When I say *prediction*, I mean a calculation that describes what is in principle the observable outcome of a real-world situation. The definition implies further that as of yet, no known classical formula or other such algorithmic calculation for reckoning the outcome exists. This is what computer scientists have in mind when they speak of machine prediction. A database isn't making a prediction, for example, when I check to see how often Ms. Jones has watched a movie on Netflix recently. That information has been stored on the computer and needs only be retrieved from its proper location. The same goes for solving a math problem—calculating which day of the week May 23, 1998, was—say, the date of Ms. Jones's last rating. It isn't a prediction in the sense specified here because we aren't dealing with an observed behavior but a calculable property.

As it turned out, predicting ratings was no walk in the park. That was because the data Netflix published didn't contain a great deal of information, looking more or less like the following:

On 6/13/2005 User 10,380 gave *Star Wars* (1977) a five-star rating.
On 10/10/2005 User 10,380 gave *Pretty Woman* (1990) a two-star rating.

We knew virtually nothing about User 10,380, in other words—just a small bit about their movie-watching preferences. Then there was the issue that for a scale, one to five stars is quite rough. Essentially, it only lets you say something like: one star = I hated it; three stars = it was fine; five stars = I loved it. Still, at the time it really was big data. One hundred million data sets: I could only store a fraction of it on my laptop![7]

Alongside the first set of data, Netflix published a second set of 2.8 million entries that looked like the following:

On 7/15/2005 User 10,380 rated *Star Wars: Episode V* (1980).

The rating itself was missing, in other words, and it was now our job to predict how many stars users had given each film. Again, the word *predict* is somewhat of a misnomer here, even if it is a standard industry term. Netflix knew how the users had rated the films in the second set, after all; the ratings had been given in the past.

Be that as it may, in computer science, anything that a machine hasn't explicitly received as information but concerns the measurable behavior of

a system or people counts as a prediction. Whether or not this behavior has already occurred or has yet to occur is beside the point.

If the predictions are good, it's a piece of cake to cobble together recommendations. The moment a user logs in to Netflix, the algorithm calculates predictions for her and all the films she hasn't seen yet. What's more, it presents them so that the film with the highest-ranked prediction is shown at the top, followed by the rest in decreasing order.

## RECOMMENDATION ALGORITHMS

We come across these sort of recommendation algorithms all the time on the internet, and they give us a first taste of how artificial intelligence works. Recommendation algorithms are what calculate the order in which your friends' posts appear on Facebook or the tweets on your Twitter feed, for example, but also the products an online store recommends—or the films Netflix suggests. In the broadest sense, they determine the sequence in which results for a given search term show up, from listings on a job platform to targeted online advertising. From this, it's already clear the kind of far-reaching consequences these algorithms have.

What's more, numerous discussions over the past several years have highlighted the likely role these same algorithms have played in making our democracies as vulnerable to attack as they are through fake news and manipulation. That's why in what follows, I would like to introduce you to a couple of these algorithms. Here too we will find the OMA principle at work: to be able to interpret an algorithm's results meaningfully, the operationalization, modeling, and algorithm must all work in concert.

What can I say? It hasn't always been like this—and for a long time, the mistakes went unnoticed. Before I come to that, though, and in order to get a feel for just how tricky predictions like these can be, I'd like to show you how data scientists initially tackle a problem.

### I KNOW WHAT YOU DID LAST SUMMER: VOICES FROM THE PAST
For the Netflix contest, it seemed an obvious choice to begin with a method that followed the person. For example, every participating team likely tried the following simplified algorithm at least once: For the data point given above, where User 10,380 rated *Star Wars: Episode V* (1980) on 7/15/2005, they calculated the *average* number of stars the user gave films.

The prediction for User 10,380, in other words, is that they will continue to behave as they have in the past—which would then allow you to predict the average number of stars other users gave *Star Wars: Episode V*. This follows the *behavioral model*, according to which all users will behave like the average.

Great stuff, really. Good enough for starters anyway. But how exactly are you then supposed to measure the quality of the resulting predictions? To do that you need to select a *quality metric*—another necessary step for any algorithm involved with machine learning. The quality metric determines the point at which we can begin to trust the decisions an algorithm makes.

### GOOD DECISION: CHOOSING A QUALITY METRIC

The quality of the prediction was measured as follows: if the algorithm predicts 3.6 stars but the user gave the film a four-star rating, that difference is squared, making it unimportant whether it was over- or undervalued. A prediction of 3.6 stars or one of 4.4, in other words, leads to the same result: $0.4 \times 0.4 = 0.16$.

It wasn't just one prediction we needed to make, though, but many. To do that, we took the average of the squared errors from all 2.8 million test data sets, then took the root of that average.[8]

With this sort of quality metric, comparing directly between various algorithms and their predictions is child's play—if you know what the actual ratings are, that is! But we didn't, of course; only Netflix did. During the competition, you could only upload your predictions for the set of test data once a day, and the system would then rate them. The company also published a list of "top scores" showing how well each team was doing at any given time—a good way of pushing us to give it another try using some other, more clever method. And try again we had to, because the simplistic methods described previously were nowhere near good enough, if not entirely off the mark. It does tend to be the case that if a user generally rates films positively, she will continue to do so in the future. By the same token, if a movie is highly rated by many users, there's a high probability the next user will like it as well. Still, those aren't exactly rules for being right 100 percent of the time!

Netflix offered a reward of $1 million for the team that designed an algorithm to best Netflix's own algorithm, Cinematch, by at least 10 percent.

The quality of Cinematch's predictions was nothing to write home about, by the way; on average it miscalculated the actual rating by nearly an entire star. More specifically, the first algorithm in the contest whose average absolute error rate lay at or beneath 0.8572 would receive the princely sum.

I'm probably speaking for most contestants when I say that while the prize money may have captured a great deal of media attention, I joined the competition for entirely different reasons. My motivation lay in proving that I was quicker on my toes than the others, that I could discover more telling clues in the data for more accurate recommendations better than the rest of the world—and, what's more, do it all as quickly and efficiently as possible! That is the real magic of data mining, or exploiting data: finding tiny treasure troves of insight that can be used down the line for making decisions.

The contest opened on October 2, 2006. It took just six days for one developer team to beat Cinematch's accuracy; a week later, three teams had managed. By the half-year mark, more than twenty thousand (!) developer teams had registered for the competition, uploading predictions for the test set more than thirteen thousand times.[9] After a year, the number of teams had doubled to forty thousand, and Cinematch's quality had improved by 8.43 percent. It wasn't until September 2009, however, that an algorithm finally beat Netflix's recommendation algorithm from three years previous, though when it did it was by more than 10 percent.

I myself toyed around with several ideas for the algorithm, but in order to win the contest you would also have had to predict which films a person *wouldn't* like. My interest lay elsewhere: Based on the small amount of data contained in the set, was it also possible to figure out which films were of a similar nature? Would users' rating behavior allow one to discover groups of similar films?

GREAT EXPECTATIONS

The original goal of the contest was to predict whether a person does or doesn't like a movie. Ultimately, however, to make a good prediction for movie night, you basically only have to understand what it is a person likes, not so much what she dislikes. With that in mind, I developed an algorithm that checked for whether two films shared an uncommonly and unexpectedly large number of fans in common. My thought was that two films many people liked in common would be well suited for a positive recommendation. Users who liked Film A will also like Film B because so many people before them have also liked both Films A and B. That might in turn point to other similarities between the films that could prove relevant for viewers. But how was I to know whether or not two films did in fact have enough fans in common to make such a recommendation?

Let's take two pairs of films, Films A and B, and Films X and Y. Out of ten thousand users total, only twenty-three people liked both Films A and B.[10] Of the same ten thousand users, however, 1,179 people liked both Films X and Y. Based on that information, is it possible to recommend Films B or Y if a new user said they liked Film A or Film X, respectively? In this scenario, you would have to determine whether the twenty-three fans Films A and B have in common and the 1,179 fans of Films X and Y contained insightful information or not about whether people that like one film will also like the other. And I've led you astray, of course: put like this, it's impossible to answer the question in the first place. To do so, you'd have to know first how popular the films themselves were. As it turns out, out of ten thousand users total, Film A received a positive rating from just forty users, while Film B had a scant seventy-three positive reviews. Hardly primetime events, in other words!

Figure 17
Can the number of people who like two films provide clues as to whether those two films are similar? If yes, then is twenty-three shared fans enough to indicate similarity? Are the 1,179 people who liked both Films X and Y a clear sign of the two films' similarity?

Films X and Y, on the other hand, were your average blockbusters. Out of ten thousand users, 4,080 people gave Film X at least four stars; 1,930 did the same for Film Y. It's not so clear anymore whether knowing that Films A and B have twenty-three fans in common tells us as much as knowing that Films X and Y share 1,179 fans.

As I mentioned before, recommendation algorithms are a sort of common ancestor to machine learning based on digital data, so it's worth taking a closer look here to understand all the levers and pulleys that go into the machinery. Here too, it quickly becomes apparent that the key issue lies in finding suitable *modeling* for the problem. Yet more than anything, the example demonstrates that with such a wide range of design options, later on it can become quite difficult to discern whether an algorithm's decisions are actually right or not. As it turns out, there are *dozens upon dozens* of ways to establish a mathematical relationship between the number of films with shared fans and all the films that have been rated or liked.[11] Once again, *operationalization* is required—that is, the question of how to measure objects' similarity.

At this point, however, the natural scientist in me reared her head, and I began to wonder whether we couldn't use a test for statistical significance. Could you assess the number of shared fans statistically, as I had done with the survival rate for my yeast cells? Such tests look to determine the frequency with which something appears *by chance*, even if no deeper connection exists. To do so, you need a control group. That was straightforward in the case of yeast cells, but what would the control group look like in this case? Well, to begin with my team and I posited that there were groups of films that many groups of people liked to watch. It's an everyday experience: if your friend likes one film with Johnny Depp in it, chances are that he or she will like other films with Johnny Depp. In the case of the Johnny Depp fan, the reaction to the two films would be linked. Yet what if no such connection existed? How often would people simply happen to like both films by coincidence? The difference between the number of fans we expected to find who happened to like both films by coincidence, on the one hand, and the actual number of common fans observed in the real world, on the other, would let us draw conclusions about whether the films' content was in fact somehow related. The more the observed number of common fans exceeded expectations, the greater the *statistical significance* and the safer it was to assume we were dealing with two films that people would always like as a pair.

To calculate how many people we could expect to like both films purely by chance, we reshuffled the film ratings entirely without regard to actual preference. This "random" reshuffling of ratings can be modeled in different ways; coincidence, it turns out, does not always happen by coincidence! The simplest model—one that has been highly popular for over thirty years now—only considers the films' popularity. If 70 percent of real-world users like the film, the "random evaluation model" randomly associates it with 70 percent of people. This is then done for each subsequent film until you have the same number of ratings as in the actual data set—but this time without any preferences for specific films.

You repeat this process a thousand times, say, counting in each case how many fans Films A and B share under those conditions. If both films are highly rated overall, they could obviously still have a number of fans in common even after the ratings have been randomly redistributed simply because they are so popular. But if it is highly unlikely for so many common fans to be observed as such, this bolsters the suspicion that the number isn't merely due to the films' popularity. That in turn supports the hypothesis that the finding is statistically significant, and thus that people who like both films do so because they resemble each other in content.

This random evaluation model holds the tremendous advantage of allowing you to calculate directly how many fans two given films will share on average. The calculation rests on the fact that two things are occurring independently of each other: in order for Films A and B to have a common fan, the same user must be randomly selected for both Films A and B. The math is easy: you multiply the probabilities of each.

It's quite straightforward to explain why that is. On a trip to Italy once, I found myself in dire need of shoes. I ducked into a store which unfortunately didn't have the shoes sorted by size, so I had to do some rummaging. On average, I only liked every fifth pair of shoes I saw, or around 20 percent. At the same time, I noticed that only one out of every ten pairs of shoes was in my size—independently, mind you, of whether I liked the model or not.

That meant there was a 10 percent chance that of the 20 percent of good-looking shoes in the store, one pair also happened to be a size 8. In order to calculate the likelihood that the store has a good-looking size 8 shoe in

Find the
pumps in the
shoestack.

stock, you multiply the two probabilities, for a grand total of 2 percent of all shoes. That number, however, only gives us the probability that any given pair of shoes in the store satisfies both conditions: I like it and it fits. To figure out how many shoes that might actually be, you would then have to multiply the product by the total number of women's shoes in the store.

In a store with two hundred shoe models, that made four pairs total. Long story short: after a quick calculation, I left the store without any shoes—the chances were too slim. It goes to show that you do learn math for the real world, not just for school!

Things work exactly the same with a random evaluation model. You don't have to mess around with endless experimentation but can directly calculate the number of fans you can expect two films to have in common. Let's try that out now for the ten thousand hypothetical viewers we've pretended have no preferences. In the real data set, forty out of ten thousand customers total liked Film A, while seventy-three liked Film B. Using the formula of multiplying individual probabilities given above, the likelihood a hypothetical viewer will like both Films A and B comes out to a whopping 0.0000292, or $40 \times 73/10{,}000^2$. Out of ten thousand viewers, we would thus expect a total of 0.292 people to like both films—practically no one. In reality, however, the videos were both liked by twenty-three people, a number many times greater than what we would have expected. In this case, comparing the observed numbers to our model gives a clear answer: statistically significant!

Now it's finally time for the grand reveal—which pair of videos have I been talking about? You might have heard of them before, think little animated vegetables spinning moral yarns . . . Made your guess? It's two episodes from the series *VeggieTales*. How 'bout them apples?[12] The interesting thing about it is that we actually found the two films without knowing anything about them other than the number of people who liked both. This is exactly the sort of thing that sends computer scientists over the moon: discovering patterns within chaotic piles of data with as little effort as possible. Wunderbar!

And what about Films X and Y? I'll let one cat out of the bag straight-away: Film Y is *Star Wars: Episode V*. What about Film X, what could it be? First, I'll calculate how many fans you could expect the two films to have in common. Film X had 4,080 positive ratings, making it one of the absolute best-sellers on Netflix. Film Y (*Star Wars: Episode V*) was another top-rated film, with 1,930. If 4,080 randomly selected people out of ten thousand users total like Film X and 1,930 like Film Y, we'd expect 787 people to like both films. In the real data set, it was 1,179, however—or 392 more than expected! Proceeding thus with ironclad logic, we conclude that whoever likes *Star Wars: Episode V* must also love Film X. Your attention please, ladies and gentlemen, as we roll out the red carpet for a very . . . *Pretty Woman*!!!

Figure 18
Many recommendation algorithms find that whoever liked *Pretty Woman* will also like *Star Wars V*. Go, Yoda Roberts, go!

*Pretty Woman*? The one with Julia Roberts? Without any spaceships whatsoever? Richard Gere, yadda yadda yadda. Just imagine if I used the algorithm to surprise my husband: "Oh daaar-ling? Aren't you and the guys meeting up for *Star Wars* Club tonight? I brought something home for you to watch! You'll love it, my algorithm says so. But no peeking, it's a surprise." A surprise, to say the least![13]

For a good long while, this algorithm and variations based on it were quite popular, and any anomalies in content were simply "explained away." In our case, a developer would likely have argued that the confusion over that whole Yoda Roberts thing was because the algorithm had in fact discovered that as Hollywood blockbusters, both films attracted an audience for whom a film's content didn't matter all that much. The main thing was that everyone else saw it and thought it was great. Sound logical?

One variation of the algorithm (from now on, let's call it the Yoda-Roberts algorithm) was applied to supermarket purchases, with results that led to the famous beer and diapers paradox. Unexpectedly, it was discovered that at certain hours of the day, significantly larger quantities of diapers and beer were sold together. Naturally, a straightforward explanation for this phenomenon was quickly found: "Imagine you're the father of a young family. Your wife and darling child have the house in a total state, and she just bawled you out over the phone for not exactly being the world's greatest help. In a fury, you drive to the nearest supermarket, grab a package of diapers—and then a six-pack to take the edge off." In this or similar ways it was thus "explained" why this seeming paradox in fact provided wondrous insights into the shopping patterns of young parents—and why palettes of beer were soon found stacked alongside the Pampers.

I myself didn't find the algorithms' results to be intuitive when I applied the Yoda-Roberts algorithm to the Netflix data in 2007. The *Star Wars*—*Pretty Woman* thing left me feeling restless. I couldn't believe it was a true, real finding from the data. So I went back and refined the underlying concept—which at that point was twenty years old—and made a couple of adjustments to the random evaluation model that was based on it. The Yoda-Roberts algorithm implicitly assumed that every customer stood an equal chance of liking a film. This led to a side effect whereby all viewers liked the same number of films, more or less. If 80 percent of users in the model rate the first film positively, 25 percent rate the second film positively, 30 percent the third film, and so on, in the end every user will on average rate around the same number of films positively. That wasn't the case in the real world, however, as the data from Netflix showed. Most users had in fact only rated a couple of films, while others had rated thousands, even tens of thousands of movies! As it turns out, it makes a tremendous difference whether or not you incorporate this property from the actual data set into your model. It's similar to the example of train schedules, where the respective departure and arrival times had to be figured in for the results to be interpretable. In this case it wasn't only the popularity of individual films that had to be modeled, but the user's "rating activity."

The new modeling took both into account. I distributed the ratings in a way that preserved the number of ratings a viewer gave (their rating activity) as well as the popularity of each individual film. Doing so was quite a technical endeavor, with calculations that were much more involved than in the first model. The expectation, however, was that our efforts would be rewarded by an improved model. Sure enough, in a boon to long-term relationships, one first success was that *Pretty Woman* was no longer recommended as a perfect match for a *Star Wars* viewing party.[14]

For both models, the overall approach itself stayed the same, by the way: we took the real-world data for the number of fans each pair of films had in common, then compared that to the number of shared fans the model expected. We interpreted a large difference as a high probability that other users would also like both films. For any given film, then (e.g., *Star Wars V*), we could sort every other film according to that difference. Both models had their appeal: the first was simple and easy to calculate; the second stuck more closely to reality. But which was better? To decide, we needed a way of evaluating them. If we only ever looked at individual results and concocted stories after the fact about why this or that result might be right, we would never figure out which algorithm is actually the best based on the two different models. Finding the answer required a method that compared results to known facts.

## QUALITY CONTROL

By now, it was 2009. I was at the University of Heidelberg, and my students Emőke-Ágnes Horvát and Andreas Spitzer had joined in the research. Based on the new concept for customer rating behavior, we now wanted to know which model was "better" at sorting, so that for each film those that were most similar would be at the top of the list. That's not something you could say directly for every film, of course. In the first place, no single person had seen every film on Netflix, so it would have been impossible to find out which was the most similar. Second, even if there were viewers who had seen all the films, they would likely disagree among themselves.

Think back to *VeggieTales* for a moment, however, the funny series with the talking vegetables. In that case, the algorithm had actually managed to discover that the episodes had far more fans in common than if it were simply left to chance. To determine whether the algorithm did that well in general, we now wanted to sort the number of common fans each pair of films had in terms of how significantly that number deviated from what chance would predict. If our idea about shared fans worked for other films as well, when it came to TV shows with multiple seasons or film franchises like *Star Wars*, the *James Bond* films, or *Naked Gun*, the algorithm should at least sort the other films in the same series high up on the list as these were unquestionably the most similar films.

The algorithm itself doesn't have any information about whether a film belongs to a series, of course, or, if so, what series that is. The only information it receives is which viewer liked which films, nothing more. No information about series, genre—zilch. If the algorithm still manages to place other films or episodes in the series at the top of the list, that tells us two things:

1. Customer reviews do in fact contain information about which films are similar.
2. The algorithm is able to make use of this information.

How interesting!

For one portion of the films in the Netflix data set, then, we had another set of films that we thought a good algorithm would rank at the top of the list. This meant we had something against which we could evaluate the algorithm's predictions, just as Netflix had compared developers' predictions to actual ratings during the contest. In computer science, this kind of *real-world observation* to which an algorithm's results can be *compared* is called a *ground truth*. A ground truth on its own, however, still isn't enough. The method by which each individual prediction's deviation is measured and aggregated must also be *operationalized* (made quantifiable). A *quality metric* must be defined, in other words. The quality metric determines how much each kind of deviation is penalized, so to speak, and the threshold at which a prediction is considered good enough. At the time, out of any number of possible metrics, we opted for the following: if a film is one of a three-part (or four- or five- or however many part) series, we check to see how many other parts of the series are listed within the top three (or four or five or however many) recommendations.

As a concrete example, let's take a seven-season series that is generally beloved among computer scientists: *Star Trek: The Next Generation*. Working off of our ground truth, for each season of the show, we want the algorithm to assign the highest ranking to the other six seasons.

That means that if our algorithm sorts all 17,769 films according to season 1 of *Star Trek* and puts seasons 3, 7, 6, and 5 at the top of the list but not seasons 2 or 4, it has sorted four out of six correctly, or two-thirds.

Figure 19
Rating an algorithmic recommendation system. It receives a film as its input and then sorts all the other films. As described in the main text, multiple variations can exist for the same operation, each following their own logic. When assessing quality, only the top entries for every film are considered.

Is this the only quality metric available? Certainly not. Our metric did not take the sequence of the show's seasons into account when sorting, for example, but we might just as easily have conceived of a metric that did. Instead of looking only at the top rankings, we could have measured the average ranking of the other films in the series or looked at what occupied the last slot for a given season or episode. I remember trying out a couple of these ideas, in fact. Many were unsuitable because it proved difficult to compare and add up values for series with different numbers of episodes. That's an understandable reason for not using a quality metric. Ultimately, there isn't a justification for everything, however, and the team simply decides which metric to use.

This explains why operationalizing one's concept of quality—that is, determining how you measure it—must be a transparent process. It is the only way other teams will be able to reproduce your results or—if they disagree with the quality metric you've chosen—use a different one to make new calculations, then compare.

One related aspect is that our quality metric would have remained unswayed if our algorithm had introduced other series from the *Star Trek* universe into the mix—something from *Voyager* or *Deep Space Nine*, for example. That is something we would ignore even though it makes sense in terms of content because our ground truth had established that only seasons from the same series (*The Next Generation*) counted. Evaluating an algorithm, then, depends on the ground truth and the quality metric selected—*both of which are defined by individual and subjective decisions.*

Once both the test data set and the proposed quality metric have been accepted, the quality of the respective model can be evaluated. And using this method, we were in fact able to show that our model was better suited to identifying similar films.[15]

Assessing the quality of a set of results, then, is used to determine which model results in the most accurate relevance ratings vis-à-vis the "truth" we've defined as our desired outcome, or ground truth. It's a bit like consulting with various astrologers, soothsayers, visionaries, and economic gurus about which stocks will bring the greatest dividends over the next five years, following all their recommendations, then in five years looking at each of your accounts and subsequently following the astrologer, soothsayer, visionary, or economic guru that brought you the fattest paycheck (or smallest loss?).

As for Yoda Roberts, my model no longer sees any reason to recommend *Pretty Woman* to *Star Wars* fans. I also have a strong suspicion that the beer and diaper paradox wouldn't show up with our model, either. I would consider it an *artifact*—a result artificially brought into creation by people—for which the random evaluation model alone is to blame. Unfortunately, we can't test it because the data from the original study isn't available.

### POST HOC EXPLANATIONS

One important aspect of recommendation algorithms is that in truth, they aren't algorithms. They are heuristics. Remember the difference: algorithms can be proven to calculate a result that solves the problem. Heuristics, meanwhile, try to find a solution based on a cost function that may be good, but for which nobody knows how far it lies from optimal. While we might stick to the term *recommendation algorithm* in what comes as it would be used in most media, it's important to understand how it is created: Data scientists prescribe what the solution for one part of the inputs should look like (the *ground truth*) and how to measure whether a solution approaches this ground truth (*quality metric*). We then select the algorithm that comes closest. More than any other lesson, working on the recommendation algorithm taught me one thing: so long as there is no ground truth deciding what the correct answer might be and an algorithm's results appear in the least bit reasonable, humans will be able to come up with a story that explains the results.[16] That's how both the Yoda Roberts phenomenon ("But *Pretty Woman* and *Star Wars: Episode V* are both Hollywood blockbusters!") and the beer and diapers paradox were long explained away. This may well be an expression of human creativity, but it is actually quite a negative trait as it means we have poor intuition about what "the right" results from data should look like. "Somewhat plausible" is more than enough to convince us as humans. But it's not exactly an ideal starting point for an artificial intelligence tasked with finding the next generation of successful employees, say, or predicting the recidivism rates of convicted criminals. Here too, it's difficult to evaluate the data at a glance.

This makes it all the more important to have a ground truth against which we can test the computer's results. The ground truth also has to add up—it must be valid, representative, and not contain any forms of discrimination. The quality metric then allows us to measure (operationalize) a given answer's relative proximity to the ideal prescribed by the ground truth. It's what anyone does when they are grading, by the way; they gauge how far off the student's answer lies from the ideal or how much of the ideal the answer contains and grade accordingly.

In many situations, the quality metric exists within a very narrow range of options. If a heuristic calculates a route from A to B, for example, and we then compare that answer to the shortest route, the difference between the two lengths presents itself as the obvious choice in a quality metric. Things are much more complicated when it comes to predicting successful job applicants. Should the ground truth consist of two groups, successful and unsuccessful applicants? Or should all applicants instead be listed in decreasing order of success, with the "most successful" at the top? And how are we supposed to evaluate the difference between the machine's answer and the ground truth we've selected? In this case and others like it, there exists a significant amount of leeway for decisions, yet they aren't decisions that are really technical in nature. And here is where data scientists, unions, employers, and other players both can and should search for common answers.

This has all started to sound as though it might be pretty complicated to get right. But why then does big data succeed as often as it does? Well, in principle it only has to be more successful or efficient than people would be in a given situation. The vast majority of recommendation algorithms belong to the kind of systems that in most cases work quite well.

## ARE WE ALL THAT PREDICTABLE?

Human group behavior does in fact often allow for very specific and well-tailored recommendations, something that never ceases to amaze my audiences when I give talks. It's also easy to understand the surprise from the perspective of an individual person. Most of us are convinced our behavior is highly individual, though even that is inconsistent and unpredictable. How, then, is the sort of recommendation system I'm describing capable of making good predictions based on the erratic and idiosyncratic decisions of others?

In the case of viewing recommendations, there's also the fact to consider that many accounts—on Netflix, but also Amazon or YouTube—were at first explicitly designed for entire families. In the original data for the

Netflix contest, there actually arose a situation where an inordinate number of accounts had given both *Sex and the City* and children's films positive ratings. This presumably had to do with the fact that mothers were using the same Netflix account as their children. The companies have responded in the meantime, of course; to obtain viewing data with greater differentiation, they now allow users to create multiple subaccounts within one paid account. If you've ever asked yourself why your Netflix account allows more than one user, there's your answer. Dividing the account up like this allows improved recommendations for each individual user, but it especially helps Netflix itself, which uses it to further refine its recommendations.[17] As far as my own viewing is concerned, I'm pleased as punch that neither my husband's manga films nor the umpteenth episode of *SpongeBob* show up in my suggestions. For his part, my husband would likely be pretty underwhelmed if he had to scroll past all my recommended cooking shows and documentaries.

Here too, then, the data—the film ratings from any one account—is noisy. Much like a staticky image on the TV, it contains both meaningful information and chance elements. The same phenomenon crops up in other sources of big data. A person's shopping behavior on an online platform, for example, often contains information that doesn't really tell us anything about them in particular. People regularly shop not only for themselves but also for a friend or relative, in the same way they might look something up on the internet for someone else. The actual pattern—the products that I personally like—thus becomes overlaid with products that my friends and relatives like. Anyone who's ever ordered something for their great-aunt Hattie only to be driven insane by the recommendations that follow knows what I'm talking about ("No, I really *don't* want any floral knickers, *thanks*! And yes, I've already got my prune juice supply squared away for the coming year!").[18]

Despite imperfect data and individual patterns, it really is still possible to derive high-quality recommendations that contribute an enormous amount to online stores' profit margin. Why is that? Well, I hate to break it to you, but in one part it's because in real life we behave nowhere near as uniquely as we'd like to think. Our individuality tends to lie in the specific combination of our individual interests, not the interests themselves.

Taking myself as an example, I tend to buy books about computer science and scientific theory, and fairly idiotic romance novels with the following formula: She finds Mr. Right, he turns into Mr. Wrong, but there's a happy ending, preferably with a whodunit story wrapped in to boot. Now, the "computer science book + romance novel" pairing isn't one you're likely to find among other customers all that often. A large number of customers, on the other hand, will have bought the same two books on computer science as me, while others will have bought the same two romance novels. Out of the undifferentiated mass of total purchases, this makes it possible to pick out patterns that do in fact appear frequently enough to make it highly probable they didn't arise simply by chance. Within the group of people that buys a certain kind of book about computer science, this makes my shopping behavior predictable to an extent. By the same token, my behavior is also somewhat predictable within the group of people that buys a certain kind of romance novel. It's the combination of interests from all the different corners of my life that first makes me an individual. Mistakes happen, of course. Thus, a search on Amazon for Uwe Schöning's absolute classic *Theoretische Informatik—kurz gefasst* (Theoretical computer science in brief) also brings up the fourth, most recent volume in the *Red Rising* science fiction series under the "Customers who bought this item also bought" section.[19] That's still somewhat plausible, as almost every computer scientist I know does in fact read fantasy and science fiction novels. How in the heck Dieter Burdoff's *Introduction to Poetic Analysis* (in large print, no less!) also wound up there is something only Amazon's recommendation algorithm could tell you.[20]

Aside from the fact that we're not nearly as individual as we often feel, the second important thing to realize is that computers are not magically plucking certain groups or categories out of thin air that simply happen to match our own interests. As it turns out, things work the other way around; products around certain subject matter exist in the first place *because* groups of people are interested in them. The wide range of romance novels and computer science books out there is explained by the presence of sufficiently large groups of people who are interested in and willing to purchase books in those subject areas. Their behavior gives rise to the choice in products, leading those same people continue to shop within the group, thereby reinforcing the pattern. These are the patterns recommendation algorithms then detect using a dizzying array of methods and reflect back. By no means are algorithms discovering behaviors nobody knew about beforehand; the producers already knew that a market existed and were responding. As consumers, we are only reinforcing the pattern(s) producers have recognized.

Recommendation systems, then, are classical applications that rely on big data. They also belong definitively to the broad class of methods of machine learning. These methods distinguish themselves by drawing conclusions based on previous data and storing them in a structure that is suitable for making decisions about future data. Every method of machine learning in turn belongs to the broader field of artificial intelligence. Does that make recommendation systems truly intelligent, though?

At least in the early days of online shopping, hardly anyone could have imagined these systems posing a threat to salespeople in brick-and-mortar stores. Especially when it came to specialty shops, a store clerk was considered the only source for advice that was tailored individually to the customer. All that changed when the technologies considered here were introduced. Most of us were pleasantly surprised by the quality of the recommendations that online stores managed to provide in time. It was astounding that something as "noisy" as the sum total of all the products a person had ever purchased could contain so much meaningful information. I myself can definitely recall a time when I was reading so much that neither my favorite bookstore in Tübingen, Osiander, nor even my favorite bookstore clerk could keep up. It was only on Amazon that I was able to find the latest books, thanks to their ingenious "Customers who bought this item also bought" function. Today, on the other hand, I don't often find the recommendations on Amazon very helpful when I'm looking for something entirely different. All the tempting new releases are often buried beneath a bunch of useless stuff and divided up into groups that make no sense whatsoever. What I wouldn't give to get my hands on the data and see whether one of our methods couldn't do better! Thankfully, one can still always pay a visit to actual bookstores, many of which house a tremendous amount of specialized knowledge and riveting finds that go far beyond what algorithms could recommend. It's why I'll sometimes book my train connections through Frankfurt with an hour to spare, so I can visit the exceptional psychology/IT/economics department in the Schmitt & Hahn bookshop. That, or I will quietly slip away between meetings in Berlin for a trip to the Dussman bookstore, to emerge a short while later laden down with heavy bags and a host of new ideas.

Seeking out patterns in data that doesn't seem to offer much by way of insight, however, is exactly what artificial intelligence does. It's how researchers in the field try to come up with new ways for computers to accomplish tasks that to date have been reserved for humans. Once they succeed, it no longer seems unthinkable that machines would ever be capable of such a thing. Toby Walsh puts it as follows in his book *Machines That Think*: "For those of us working on AI, the fact that these technologies fall out of sight means success. Ultimately, AI will be like electricity . . . [it] will likewise become an essential but invisible component of all our lives."[21]

## UNETHICAL DATA COLLECTION

One problem with big data is that all this highly fascinating information lies in the hands of a small number of companies, often kept behind high walls. Another problem is that many a large data set is simply too easy to collect. Vast numbers of websites can be saved from the web, for example, using a *crawler*, a small piece of software that automatically surfs the World Wide Web. Other online services called *interfaces* similarly provide access to this kind of data en masse through databases. The simple fact that it is so easy to generate vast quantities of data makes it tempting to do just that, even without asking permission of the original creator or involved parties. In 2017, one researcher was able to download forty thousand images from the dating platform Tinder using an interface and then made them publicly available online.[22]

You may be asking yourself why such interfaces exist in the first place. The answer is at least twofold, and the first part is highly pragmatic. In principle, anyone who signs up with Tinder can call up and save the same data manually; an interface just makes it easier to do. Second, interfaces allow other software providers to develop services used on Tinder, giving rise to a sort of digital ecosystem around platforms that can be useful both for the original service provider (Tinder, in this case) and users alike.

The first argument—that an interface merely simplifies an action that can be done manually—is not entirely valid, of course. There's quite a difference, after all, between someone going to the trouble of picking out and saving forty thousand photos by hand and a digital aid doing the same thing. And it is only in the former case that our gut tells us it isn't so bad if such data is shared publicly. At the individual level, a person would only look at that much data if there were a good reason to do so—but there isn't, so no one does. This explains further why users often feel as safe as they do and that they "have nothing to hide." They unconsciously calculate the amount of time and energy it would take a real person to spy on someone, compare it with how interesting they themselves are, and conclude, "Who would do that? And even if they did—I have nothing to hide."

That cost-benefit analysis changes radically, however, once an interface makes it a piece of cake to capture and search through large quantities of data digitally. It also makes it much more straightforward to simply go ahead and see whether it's possible to obtain information from forty thousand photos and, if so, how much. Suddenly, everyone is interesting.

The published Tinder photos haven't been the only scandal surrounding big data—not by a long shot. In 2016, Danish researchers published a similar data set they'd lifted from another dating website; this time, it was seventy thousand records containing users' personal data. The moral outrage was universal—and the data scientist responsible was baffled by the uproar. This brings to light another aspect of the world of computer science: sometimes the rationalist mindset carries the day. The data scientist defended himself by explaining that "all the data found in the dataset are or were already publicly available, so releasing this dataset merely presents it [in] a more useful form."[23]

Sometimes data collection occurs by other means. In January 2019, the hashtag #10yearchallenge went viral on Facebook. Users were "challenged" to post a current profile photo alongside a picture of themselves from ten years ago, and a pox on anyone who thinks ill of it. That included people like tech writer and author Kate O'Neil, who in response to the trend tweeted: "Me 10 years ago: probably would have played along the profile picture aging meme going around on Facebook and Instagram. Me now: ponders how all this data could be mined to train facial recognition algorithms on age progression and age recognition."[24]

In an article for *WIRED* magazine, O'Neil works out in detail how it would already be quite straightforward for Facebook to compare profile pictures with a ten-year gap. Yet for purposes of AI training, participants' data from the ten-year challenge would likely be far more reliable.[25] After all, O'Neil writes, a lot of people also use (or have used) a picture of their dog, their kids, or some other offbeat image as their profile photo.[26] It remains a mystery to this day who came up with the hashtag and for what purpose exactly. Whatever the case may be, in an era of data collection, data sharing, and an untold number of ways of analyzing it, computer scientists have grown quite leery of such hashtags, along with other viral activities that could lead to huge data sets with a specific context that might be exploitable.[27]

In the following example of unethical data collection, by contrast, things weren't ambiguous in the slightest. In this case videos of people who had undergone hormonal gender reassignment were used as a particularly challenging set of entries for various facial recognition software. One of the scientific studies discussing it fails even to mention which dataset was used. Nor for that matter is any indication given of whether the data was collected with the permission of the people shown; at any rate, the photographic stills from the videos do not show copyright signs or signal the copyright holder's consent.[28] Writing for *The Verge* in 2017, James Vincent noted that researchers had secured the data first, only later stopping to ask themselves whether it was right to do so.[29] Vincent also reported on a conversation with one researcher, who for a while shared a collection of the videos with other researchers as a list of links. When the collection fell subject to controversy on Twitter years later, the scientist was surprised. It was only a list of links, after all; he had never used them commercially and had stopped sharing it over three years ago. Besides, he had, "as a courtesy," asked the people whether they were alright with it—though anyone who didn't reply may have ended up in the dataset anyway.

Here, too, researchers seem to have thought: "What's all the fuss? The data were public; all I did was piece them together digitally." And from the perspective of a guileless researcher, that may be understandable. I still recall how early on in my doctoral program, I considered my own attempt at a similar kind of data collection with a popular dating platform. The platform's terms and conditions hadn't explicitly banned data collection, and I wasn't planning anything evil; I just wanted to see how people were linked to each other. I sent an email to the platform indicating that all this led me to believe I had permission to read and collect any and all information available on the company's publicly viewable sites. It resulted in an irate phone call from the company's managing director to my thesis advisor. The project fell to the wayside—something that today I'm happy to report. Unfortunately, the fact that this all might look quite different to the people it affects—like losing control over their private lives—really can get lost amid the thrill of assembling data and a clear conscience about one's drive to research.

In the meantime, all these experiences have left me somewhat paranoid. Recently while out shopping with my nine-year-old daughter, I asked her to look up a store's opening hours on Google. No sooner had the words "Hey Google, when is" escaped her lips then I leapt out of the driver's seat and cried, "We don't speak with Google!" She shrank into a tiny ball, as did I, before excusing myself many times over. Of course, she couldn't have known; she simply saw the microphone symbol and used it. I also know that my reluctance is both outdated and irrational; voice input is the way of the future. Using the keyboard and mouse as interfaces for issuing commands to a computer was always more of a crutch than something to be taken seriously. Still, I don't want to simply hand my voice over without putting up some sort of a fight. Anyone who has seen how deepfakes can be created from videos and speech samples will know why. In mid-April 2018, a video of a speech by Barack Obama gave a sobering example of just how easy it is today to create a convincing video in which household figures say whatever the video's creator would like them to.[30]

These days, the nontransparent and at times unethical way in which data is collected for training artificial intelligence has led me to reply "opt out" when I hear an automated voice inform me that "the following phone conversation may be recorded for training purposes, please say if you wish to opt out." As long as I still thought that it was really about training human employees, I was happy to do so. As for their computerized coworkers, I only want them trained if I know exactly what is being done with my recorded voice.

## ALGORITHM DESIGNERS' RESPONSIBILITY
## FOR THE RESULTS OF DATA MINING

So who's responsible for the results of an algorithm that runs on big data? Well, in terms of the result itself—the calculated number—I as the algorithm's designer hold clear responsibility. If I program an algorithm to analyze Netflix usage, for example, it should do what I claim it does. In other words, I'm responsible for the algorithm's programming *being correct*. If, however, someone else then turns around and uses the algorithm for a different data set, to what extent am I still liable for the results, in terms of their interpretation?

Often in such cases, there is someone else involved—a *data scientist*.[31] A data scientist is well-versed in all kinds of methods of evaluation and usually in how to visualize results. One article describes them as "the people who understand how to fish out answers to important business questions from today's tsunami of unstructured information."[32]

On the one hand, *data scientist* is essentially a new professional moniker that simply sounds more hip then good old *statistician*. Yet a data scientist does in fact bring a different set of goals to bear when evaluating data. If statistics aims more at description, data science is geared more toward discovering new patterns in data. A data scientist is also expected to be an active programmer and to be able to communicate the results of her analysis visually.

With my own work, you get both at once—I develop algorithms and apply them to real-world questions. In addition to the Netflix dataset, my coauthors and I used our recommendation algorithm on two other sets of questions. The first was in conjunction with Professor Thorsten Stoeck, a colleague at the Technical University of Kaiserslautern (TU Kaiserslautern). In the course of his work on how to describe biodiversity and its different manifestations in various environments, Stoeck gained access to a particularly fascinating set of data: samples taken from spots throughout the world's seven oceans as part of a years-long effort to learn more about the local microorganisms.

It now fell to us to see whether the distribution of those organisms was coincidental or whether ecological niches might not also be involved. One older ecological theory, global dispersal theory, states that the microorganisms involved are so small that their distribution is determined by the oceans' currents. As it turns out, we were able to show that a significantly higher number of the creatures appeared together at certain testing sites than others and thus prove that ecological niches did exist for these microorganisms.[33]

My favorite use of the algorithm, however, brought me back to my days as a biochemist. Working alongside Özgür Sahin and David Zhang at Heidelberg's German Cancer Research Center and drawing on data from a long list of coauthors, my PhD student and I sought out a new method for stopping a particularly lethal form of breast cancer.[34] To do so, we tested 850 different small molecules known as microRNA that appear in cells for their effects on the protein of the cancer. Happily, using one variant of the original recommendation algorithm, we were in fact able to identify ten microRNA molecules, three of which were able to stop the cancer in the laboratory.

These three examples—the Netflix data set, the distribution of micro-organisms in oceans, and identifying new cures for a lethal form of breast cancer—make it clear just how broad the potential range of uses for a single algorithm can be. As I've already mentioned, classical algorithms' true superpower lies in how many different contexts they can be applied to. By the same token, it means that if a designer has no control over the context of an algorithm's application, he or she cannot be held responsible for the interpretability of the results. In the case of the Netflix algorithm, for example, our first random evaluation model could only be used under certain conditions; otherwise, we'd get wrapped up in the beer and diapers paradox. If I'm aware as the designer that the data are only interpretable under certain conditions, I have a responsibility to communicate that. I could even design a small test for monitoring purposes, in the form of a message that flashes up on the screen: "Hey user, your data doesn't match the model—are you sure that you want to use it anyway? Your results' interpretability may suffer! Click on YES or NO, and if you're stumped, simply dial our help line toll free at 1–800-ALGODOC!" I certainly could have used a message like that myself over the course of my biochemistry degree, for the times I ran up against the question of how and under what circumstances one analytical method or the other could be used. It's also why today I view developers' main responsibility to be communicating about exactly which assumptions lie behind an algorithm. As the designer, however, I can hardly take responsibility for who winds up using my algorithm on their data, nor for that matter the conclusions he or she may draw from its use. The range of possible applications—and that includes inappropriate applications—is simply too broad.

A further point seems important in this context, which is that inspecting an algorithm purely in terms of its coding won't yield much. Of course, errors in implementation will occur from time to time, but their analysis and repair are exactly the part of the puzzle that computer scientists are well-equipped to solve. The sort of algorithmic safety administration that Viktor Mayer-Schöneberger and Kenneth Cukier propose would only examine an algorithm's code for its functionality, which means it wouldn't be of any further assistance when it came to determining whether or not the algorithm was being used in a sensible way. Yet the truly serious errors occur when *interpreting an algorithm's results*: that is the single most important lesson you can draw from this chapter. The OMA principle leads to problems when it isn't followed. It's only when each and every modeling step and operationalization (making social concepts measurable) has been finely attuned to the algorithm that the results will be interpretable. But when computer scientists go to process data from social interactions, we lack a working knowledge of the social concepts necessary to model and operationalize algorithms because they weren't included in our training. Nor, I would like to emphasize, is it truly helpful or sufficient to teach these aspects to computer scientists in great breadth and depth. The first problem is that computer science already encompasses a very broad field of study—and necessarily so. Any new content would have to fight for room and replace something else that's been deemed basic knowledge. Second, as the algoscope has shown, only a relatively small number of software systems truly merit scrutiny. Our solution at the Technical University in Kaiserslautern was to introduce socioinformatics as a new course of study that would combine computer science with other disciplines, and train students to recognize, model, and, if possible, predict the social effects of software. I go into this in greater detail in part III. Before that, however, here's my answer to the frequently asked question of who is responsible for the results of algorithms that work with big data:

**Who's responsible for how an algorithm is used?**

When it comes to data mining, designers often won't know which contexts their algorithms will be used in. The vast majority of basic algorithms are designed to solve abstract mathematical problems. This means designers' principal responsibility lies in clearly formulating how the underlying mathematical problem has been defined and what conditions the data must fulfill for the result to be interpretable in the first place. It then falls to a data scientist to decide whether an algorithm can be meaningfully applied in a given context.

AT A GLANCE: BIG DATA = BIG RESPONSIBILITY

In summary, when computer scientists talk about *big data analytics*, they are talking about methods of searching for patterns among data that usually wasn't gathered with that initial purpose in mind and that may contain errors but is available in very large quantities.

In the Netflix dataset, for example, the ratings we received weren't complete. Most viewers had rated only a handful of films—far fewer than they had seen over the course of their lifetimes, at any rate. On top of that came the consideration that there was often more than one person using the same account. Overall, the data we were working with wasn't truly reliable or complete: it was noisy! Our ability to use big data meaningfully despite this owes to the fact that we possess so many sets of information, which on their own say little, and limit us to searching only for correlations. By *correlation*, I mean information stating that two things often appear together; "a customer who likes Film X will usually also like Film Y," for example. The same principle applies in online shopping, where it is often helpful to know that most if not all customers might be interested in a particular product. On the internet, even minor improvements can bring significant earnings.

In order to assess how well software based on such patterns is able to predict future behavior, we need a *ground truth*, which gives the prediction a basis for comparison. We also need a *quality metric* to tell us how the comparison is to be gauged overall. The Netflix example showed us that many different methods for making predictions can exist simultaneously; in this case, data scientists simply tried out a number of different methods and let the quality metric determine the best model, aided by a ground truth.

Another point to bear in mind—and one we will return to later—is that as long as we humans are able to find a somewhat plausible explanation, we aren't very skilled at recognizing an algorithm's mistakes for what they are. Wrong decisions or answers, however, can have a serious impact on the lives of individual people. This makes it all the more important for both the ground truth and the quality metric to be clearly defined, by an extensive process that brings as many affected parties and users to the table as possible.

A big data approach, then, makes use of large quantities of data that are in and of themselves not particularly meaningful in order to recognize statistical patterns. Statistical patterns apply only for large groups of people, not necessarily for an individual person's behavior. The next giant step in this process is machine learning, which searches for patterns within data from the past in order to then make decisions about new data. In machine learning, a set of rules is created directly on the basis of whatever correlations have been discovered and is then used to make predictions. In terms of scientific theory, the process leaps directly from formulating a hypothesis to applying it, without providing any experimental validation. It legitimizes itself by employing sets of test data with a ground truth: if a set of rules proves accurate for new data from situations whose outcome is known, then what we've learned must be right.

To form our own opinion of machine learning, in the next chapter I introduce you to the final letter in the ABCs of computer science: C, for computer intelligence.

# COMPUTER INTELLIGENCE

Today's computer scientists would most likely take quite a different approach to Donald Knuth and Michael Plass's idea of automatically improving the layout of a text by using a "beauty function" to set line breaks. As described in chapter 3, Knuth and Plass began by sketching an evaluation function designed to suit their tastes, which included a large number of parameters that could result in an aesthetically pleasing text. Only then did they develop an optimization algorithm to maximize the "beauty" of a text, using the evaluation function designed in the first step. That's how classical algorithms work: first a model of the world is sketched ("What makes the layout of a text beautiful for me?"), followed by an algorithm that tries to find the best answer according to that model.

Programmers who work with machine learning, by contrast, wouldn't bother at all with formulating what for them characterizes a well laid out text. Instead, they would first assemble a large group of texts that had been manually set by acknowledged experts to serve as the underlying data. Next, they would use algorithms to search for quantifiable properties (i.e., generate a model for "aesthetic text layout" from the data). Finally, a second algorithm would generate line breaks based on that model.

And with that, we've already come to the heart of the matter. The algorithms of artificial intelligence are intended to take over for cognitive activity, something that up to now we've regarded as quintessentially human. Others might speak of *automating intelligent behavior*. Today, computer science defines *artificial intelligence* in broad terms: as software that aids a computer in performing a cognitive activity for which humans are normally responsible. Such a definition contains multiple problems, of course. To begin with, it begs the question of what exactly cognitive activity or intelligent human behavior is. Second, the definition shifts once the goal has been reached. As soon as a computer can do what we wish it to, we perceive its accomplishment as requiring less intelligence for the very reason that a computer was able to do it. This has led Toby Walsh to speak of a moving target for artificial intelligence.

It's also necessary to distinguish between *weak* or *narrow AI* on the one hand and *strong AI* on the other. Weak AI is capable only of solving specific problems—playing chess, for example, or recognizing what lies in a picture. Strong AI, by contrast, would designate a computer system that responds intelligently at a general level, including in situations where precise factual information is missing or the objectives are unclear. By and large, computer scientists are of the opinion that *artificial intelligence* is a misnomer for the field of research and that the definition itself is so muddy as to be practically useless. As Florian Gallwitz writes in *WIRED* magazine: "If you broaden the definition, even an everyday pocket calculator meets the requirements for what is called 'weak artificial intelligence.' It seems just about as sensible to in all seriousness lump paper airplanes, New Years' Eve fireworks and tennis balls together under the title 'weak interstellar satellites.'"[1] What is clear is that for those working in the field today, the term AI brings together a wide range of technologies that tend not to be what the media has in mind when it reports on artificial intelligence. Examples of those technologies include expert-, diagnosis-, and knowledge-based systems, which store known facts and decision-making rules in such a way that they can later provide answers to certain questions. This is feasible in contexts where a limited number of decision rules can provide highly accurate results—a diagnosis system able to state which symptoms are associated with which disease, for example.

The trouble is that in most situations, humans intuitively take so many rules into account that it seems well-nigh impossible to file them all away in a structured manner. Humans, for example, would say:

"He sat down on the rug."
"He walked on the rug."

In both sentences, the word *rug* has the same meaning.

Such a rule can't be applied with the following two sentences, however:

"He sat down on the bank."
"He walked into the bank."

While the verbs are the same as in the first case, the word *bank* has a different meaning each time.

By the same token, if a person reads the sentence: "The child dropped the pan because it was hot," then she knows that the pan is hot, not the child.

For a machine to understand text in this way, the requisite knowledge would have to be entered into an expert system so as to make illogical conclusions impossible. Such systems were in fact developed over the course of the twentieth century and are still in use today. But it's turned out to be quite a headache to store all the implicit knowledge required.

Machine translation is a particular sore spot for knowledge-based methods, especially when proverbs or double entendres come into play. Just look at some of the translations in figure 20 that Artie has come up with for different businesses in the past![2]

Figure 20
Artie's attempts at translation. He can already speak (a little) English!

The quality of machine translation first improved with a radical new approach that dropped human-made rules in favor of the notion that computers could "learn" how people translated certain words if they had enough examples. To do so, machines were fed texts that existed in two languages. These are quite common in the European Parliament, for example, which translates its documents into all twenty-four of the EU's official languages. Instead of risking linguistic analysis, then, the computers translated by way of analogy.

These days, the subject heading "artificial intelligence" has come to mean anything that *imitates* cognitive ability—which includes the expert systems and knowledge databases of the 1980s. Yet it also includes the algorithms of machine learning, and especially the neural networks and "deep learning" that are such hot topics at the moment. Ahead, I briefly explain what goes into constructing neural networks, which are responsible for our current breakthroughs and are nearly always what the speaker has in mind when you hear about "artificial intelligence," "self-learning algorithms," or even plain old "algorithms."

I also bring up a couple of examples that draw attention to the need for greater differentiation between terms. As it turns out, not all algorithms learn from data. The vast majority are "classical" algorithms and thus controllable to an extent. What's more, the algorithms that do "learn" from data are themselves static; they don't change. But first, here's a somewhat more specific definition for you to mull over:

> When the media talks about AI or (self-learning) algorithms, it is usually only the *methods of machine learning* that are meant.

### HOW COMPUTERS LEARN

We've now finally arrived at the million-dollar question: How exactly does a computer learn? As it turns out, it's something computer scientists learned by simply watching our kids, then applying it to the computer.

Figure 21
The spate of terms surrounding AI is fairly confusing. In a scientific context, AI is taken to refer to a range of technologies. When the media discusses AI or algorithms, they are usually thinking of the methods of machine learning.

I'll give you an example. When he was very young, my son didn't like hot food; anything past 100 degrees seemed to morph into some sort of seething witch's brew. To be on the safe side, he tended to eat his soups cold. Sooner or later, however, he noticed that soup tasted better when it was warm. In other words, he learned that soups were "safe" so long as they weren't steaming—until, that is, he burned the roof of his mouth one day on a seemingly harmless "lava soup" (cold on top, piping hot below). After that, the only thing that would help was a bit of parenting wizardry: no steam + stirring the soup three times + blowing on it three times. Abracadabra, edible soup!

In this case, my son learned through observation, constructing decision-making rules and feedback. If the feedback tallies with the rules established up to that point, the rules are reinforced; if not, they change. This is essentially what computer scientists do with computers. We provide them with information in the form of data, a structure into which the decision rules can be filed away, and—under ideal circumstances—feedback.[3]

As with children, the goal is for computers to adopt rules that are as general as possible. If my kid only ever learned to spot hot tomato soup and couldn't apply the "warning signs" to other kinds of hot soup, for example, he might spend the rest of his life obsessing over the various security clearances that his appetizers should receive. In the same way, a computer should never draw too narrow of a lesson from the small snippet of reality it's been given so that its "discoveries" can be applied to new data as easily as possible.

It's not always clear how exactly a child decides a certain situation is at hand. This became especially apparent when our family got a cat. Our son was delighted with the creature, but instead of using its actual name, Neo (or just Kitty), he called it *tst*, clicking his tongue against the roof of his mouth. We couldn't make heads or tails of it for the longest time—even though it was quite obvious what was going on. When we called for our son, it was often with the words, "Come, Fabian, let's go home!" When I called the cat, however, I said "come, *tst*, come!" Fabian had taken the clicking sound I was making for the name of the cat—a classic categorization error.[4]

This anecdote captures quite effectively how machine learning works. The algorithms of machine learning learn by way of example. Data scientists present the algorithm with various situations, then tell it how those situations are to be evaluated based on the *ground truth*—that is, the result that is to be learned. Aided by a defined set of instructions, the algorithm combs through the situations for noticeable patterns that occur regularly when the desired result appears and rarely for other results. It's the same as when we tell our children, "Now it's safe to cross the street" and "Now it's not" a million times over, in the hopes that they will learn the rules behind our statements. We don't know exactly how they do it: perhaps they learn that a street is safe when they can't hear a lot of motors, for example, or that they have to act more carefully when they see a lot of parked cars than when they have a clear view.

The patterns the computer discovers are then stored in the form of decision rules or formulas in a structure built for that purpose. This structure is called the *statistical model*, a term that, unlike *AI*, really does have something to say. It indicates that we are describing only one *slice* of reality in *abstract form*, and moreover that this description is of a *statistical nature*; that is, it doesn't lay any claim to causal relationships or hundred percent accuracy.

And that's all, folks! Here's the definition for *machine learning*:

> Automated learning by way of examples, in which decision rules are searched for then stored in a statistical model.

The statistical model itself is associated with a second algorithm that is often extremely simple in nature. It is this second algorithm that takes new data through the statistical model and the rules it contains and makes the actual decision.

When you hear talk in the media about "the" algorithm or the "power of algorithms," it is usually this second step that is being referred to: entering data into a small and simple algorithm, which then makes a decision. In the process, the workings of the first algorithm often go overlooked, which are responsible for what is actually the most interesting part of the whole affair—the statistical model. What's more, talking about the "power of algorithms" glosses over the fact that there are always people involved. Even with the first algorithm, there is no magic happening, after all, nor is it "objective" in a general sense. As for the second algorithm, which runs through the discovered rules and makes the actual decisions, it is built of nothing more than simple multiplications, additions, and if-then decisions. It doesn't need regulation or inspection by an algorithmic safety administration.

# TWO ALGORITHMS
## to rule them all

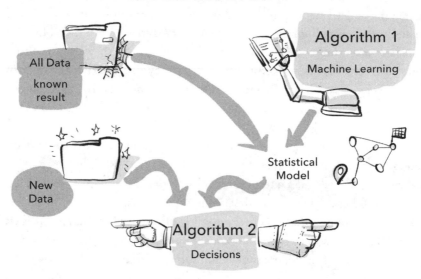

**Figure 22**
This diagram illustrates why algorithms seem to be the decisive point in decision-making, even though the more important part is learning the statistical model. The thing is that once a statistical model has been learned, a very simple algorithm can calculate a decision for new data. What is of actual interest, however, is how the statistical model itself comes into existence. It is a result of training data, the methods of machine learning, and a couple of other elements that are described in what follows.

To convince you that there's no wizardry afoot when it comes to machine learning, ahead I describe the following three methods for learning decision rules in detail:

1.   The tree of knowledge: decision trees
2.   Support vector machines—a.k.a. the stick(ler) test
3.   Neural networks (very briefly)

I then briefly sketch how algorithms and heuristics each learn these rules from data and store them in a statistical model.

To keep it simple, I use examples in which machines only have to decide between two possible results. This method is referred to as *binary classification* because it decides out of two classes, who or what belongs to which. To do so, each of the three methods described ahead constructs a statistical model that takes the shape of a decision-making structure, basing it on a ground truth that is provided in the form of a set of *training data*.

### THE TREE OF KNOWLEDGE

My husband has had a real string of bad luck driving over the past few months. First, a speed camera awarded him a pricy portrait for driving 89 km/h in what sure felt like a 100 km/h zone, but tragically turned out to be a 70 km/h. Shortly thereafter, he was caught by another, this time in a construction zone within the city limits and with equally regrettable results. In Germany, you better bet that's serious cause for concern. There's an entire agency located in the otherwise rather obscure city of Flensburg that gathers all the more serious driving infractions at a central location. Speeding or ignoring a red light: one to two points and potentially no driving for one to three months. More than eight points? Lose your driver's license for half a year at least. Whenever Germans are gathered together at a bar and there's absolutely nothing to say, you can always talk about "your points in Flensburg" to liven up the discussion. Guaranteed! We needed to know how bad it was for my husband. The decision tree shown in figure 23 can help work out how badly things may turn out for him.

As soon as we know how much my husband broke the speed limit by, we can let the decision tree run its course using a simple algorithm. Computer scientists turn the tree on its head with the roots at top out of historical convention, by the way; don't let it bother you. So we begin at the root of our upside-down tree, then proceed to answer each question that follows. Hmmm . . . it looks as though things might get hairy for my husband. If this is at least the second time he's been caught traveling 25 km/h or more over the speed limit, then he's liable to lose his license for a month, in addition to running the risk of a fine and a second point on his license in Flensburg. (A note to the reader: He was only 20 km/h above the speed limit. My husband would like to let the record state that as a precautionary measure he now only drives 50 km/h tops, both in and out of the city. Just a word to the wise if you get stuck behind a car with Kaiserslautern license plates inching along the highway.)

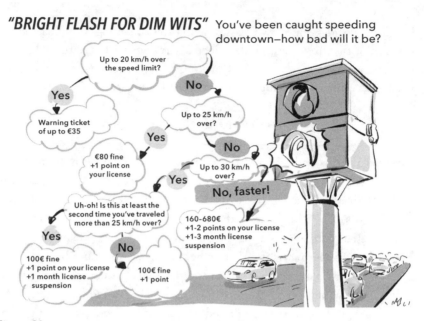

**"BRIGHT FLASH FOR DIM WITS"** You've been caught speeding downtown—how bad will it be?

Up to 20 km/h over the speed limit?

No

Yes

Warning ticket of up to €35

Up to 25 km/h over?

Yes

No

€80 fine
+1 point on your license

Up to 30 km/h over?

Yes

No, faster!

Uh-oh! Is this at least the second time you've traveled more than 25 km/h over?

Yes

No

160–680€
+1-2 points on your license
+1-3 month license suspension

100€ fine
+1 point on your license
+1 month license suspension

100€ fine
+1 point

Figure 23
The decision tree shows whether you will have to reckon with a license suspension for a speed violation within city limits.

I based the decision tree shown in figure 23 on established rules and regulations for fines, but could you also arrive at the same series of penalties and underlying decision criteria simply through observation? Of course you could—but only if you had access to a large set of data. If, for example, you had access to every fine sent out to every person caught by a speed camera, you would see that within city limits, exceeding the speed limit by up to 20 km/h only incurs a warning fine. Surpassing the speed limit by 25 km/h or more on at least two separate occasions, however, gets your license suspended for at least a month. If there were enough examples of each infraction from a given period, it would be possible to derive the rules based on observation alone (how much the speed limit was exceeded by and the consequence). It's helpful in this case that the rules carry the exact same consequences for everyone; the pattern becomes recognizable with only a handful of instances.

ALL ABOARD THE TITANIC

As it turns out, a decision tree can be automatically constructed even in cases where the decision rules don't apply 100 percent of the time, as they do with the speed camera. One classic scenario most students will encounter on their first forays into machine learning involves one of history's most fabled ships. In the scenario, the personal data of the passengers aboard the *Titanic* is given as input, and the task is to predict whether the passengers survived the trip or not. Now before you go asking why on earth you'd ever want to construct a decision tree when it's been known for well over a century who survived and who didn't—yes, of course we could simply check the list, and of course we are trying to "predict" something that lies in the past. But if it's possible to make a highly accurate prediction, that means a pattern is actually present in the data, something we can then go on to confirm using lists of actual survivors.

In this case, the problem of finding the best rules for a decision tree could be put as follows:

*Given:* A set of training data with information about some of the passengers aboard the *Titanic*, including whether or not those passengers survived the trip.

*Find:* A decision tree that makes as few errors as possible when predicting which of the rest of the passengers survived the trip.

The entire set of all passengers has been divided into two groups, in other words, and it is now the algorithm's job to discover patterns within the properties of people from the first group (the *training set*) so as to allow for largely accurate predictions about who survived within the second group (the *test set*). Once again, it's comparable to teaching children how to cross the street safely: at one point or another, a parent begins to let children decide for themselves. The more confident the children are in their decisions, the more you as a parent can also trust them to walk to school by themselves. With the computer, the test set plays the role of the parent; what data scientists look for is how well the computer assesses situations previously unknown to it. The degree of trust we place in the rules discovered by the computer grows along with the number of correct predictions.

The decision tree can make one of two predictions: survived or did not survive. There are only four possible outcomes:

1. The algorithm predicts that Person X survived, which is true. This is called a *true positive*.
2. The algorithm predicts that Person X survived, which isn't true. This is a *false positive*.

3. The algorithm predicts that Person X did not survive, which is true. This is a *true negative*.
4. The algorithm predicts that Person X did not survive, which isn't true. This is a *false negative*.

You'd be entirely right, by the way, to be confused about one set of predictions being labeled *positive* and the other *negative*. A person surviving is of course a positive event, but in this case the designation has nothing to do with whether the predicted outcome is perceived in a positive light or not. The expression comes from the medical field, which has classically described the results of a test for an infection as *positive* if the pathogen has been shown to be present and *negative* if it hasn't. For those of us who can still recall a time before HIV, we certainly would have been struck by the same odd feeling the first time we read a newspaper report about a patient being HIV positive. After all, it came at the beginning of a long and painful chapter that wasn't positive in the slightest. When we are dealing with *binary classifiers* (a classifier that distinguishes between two types of behavior or properties), one of the two decisions is simply designated positive, with it being entirely arbitrary which. Often the more "important" class is designated positive. When predicting recidivism, for example, the class or group who did recidivate would be positive; when predicting creditworthiness, it would likely be the class of people who are creditworthy.

We obviously want our algorithm to construct a decision tree that results in as many true positives and true negatives as possible. To do so, we feed the algorithm information about one portion of the passengers; a training set that in this case contains details about 891 people aboard the ship. We could just as easily have made the training set larger or smaller—data scientists usually make the call by rule of thumb. Dividing the overall data set into training sets and test sets, then, is first among the many "levers and pulleys" I've talked about that make it possible to adapt and tweak whichever method it is we're using.

Our decision tree is now constructed based on the training data.

And what form might the algorithm used to build this particular decision tree take? The basic idea is that in every step along the way, the algorithm searches for the one property among the passengers on the *Titanic* that best distinguishes those who survived from those who drowned. As we know, the data set contains a large amount of information, but it isn't complete for all passengers. The names of all 891 people are known, for example, but the age is entered only for 714. The class of ticket booked is known for everyone, as well as how much it cost. It is also known for everyone how large their travel party or family was. We know their gender and, for many, their form of address. Finally, the algorithm obviously knows who survived and who didn't—the ground truth that provides the algorithm with the necessary feedback and lets us construct the decision tree in the first place. Of the 891 passengers whose data is contained in the training set, 38.4 percent survived.

What property would you say is the algorithm likely to determine led to the best chances of survival? That's a trick question, of course, for me to determine how old you are. If you were at least twelve in 1998, then you probably saw the film *Titanic* and partook in the Passion of Leonardo di Caprio. You also know who stood the better chance of surviving: women.

As it turns out, the algorithm also finds that out of all the information listed, gender is what best divides survivors from nonsurvivors. If you divide the data set according to gender, you find the following: The training data contained 314 female passengers (35 percent of all passengers in the set), out of which nearly 75 percent survived. Out of 577 male passengers, on the other hand, only 109 survived (just under 19 percent).

The women represent a relatively homogenous group. You could of course try to break the group down further, maybe according to age or ticket class. If in the end you only wanted tiny little subgroups made up of either 100 percent survivors or 100 percent nonsurvivors, you would probably have to take passengers' names into account, too: "All women named Fatima Masselmani survived" (she did, in fact!). But would that really be a helpful rule? Clearly there was only one person aboard with that name, which makes a rule like that too specific.

As a basic rule, with machine learning you try not to overinterpret—or *overfit*, as it is called—the data set. The training set probably contains a lot of unimportant details that are better left unlearned; in terms of my son's hot soup problem, he shouldn't let himself get too distracted by the parsley on top of the tomato soup but should focus instead on the general properties of hot soups.

Yet how are data scientists supposed to determine whether a group is homogenous enough as is or should be broken down further? There are in fact more than a dozen methods for gauging the homogeneity of a given group, and just as many for deciding whether the group should be split up even further (more on that in a second). First, let's take a quick look at the group of men, which at a survival ratio of 19 to 81, or about 19 percent, is even more homogenous than the women, who survived at a rate of 75 to 25, or 75 percent. If you left it at that and simply predicted did not survive for men and survived for women, in other words, there would be fewer mistakes with men than with women. But in this case, the algorithm continues on to check whether there isn't a further division within the group of men that results almost exclusively in survivors or nonsurvivors.

I didn't write the algorithm used for the decision tree I'm describing, by the way; Wikipedia user Stephen Milborrow did, then published it online.[5] I'm basing my numbers here on his results, after taking a closer look at the set of training data he used. Milborrow names the general type of algorithm he used to create the tree, but it's one with a great number of small control levers and mechanisms for a data scientist to operate. This means I can't reconstruct why the algorithm searched for a second set of criteria to differentiate between men, but not women. This is a first valuable indication of how important it can be to transparently describe all the bells and whistles you've attached to your algorithm: it allows others to understand how you reached your results.

All we can know for certain, then, is that Milborrow's algorithm was able to optimize its results by making further distinctions. In the process, it discovered a second property that decided between life and death for male passengers: age. The group of male passengers older than nine and a half showed a lower rate of survival than the overall group of male passengers (17 percent compared to 19 percent). Within the younger group, we can see further that all eleven boys traveling in the first and second classes survived. Of the twenty-one boys traveling in the third class, meanwhile, eight survived.

There is another property that presents an even clearer picture of who survived, however, and that is the number of siblings the younger male passengers were traveling with. All eighteen boys traveling with no more than two siblings survived, but only one traveling with more than two survived. The algorithm stops at this point as the current stop criteria have been fulfilled (homogeneity and avoiding overfitting). Figure 24 shows the final decision tree as published by Milborrow.

How is such a tree, which doesn't describe rules that apply 100 percent of the time, now put to use? The parts of the tree beyond which no further

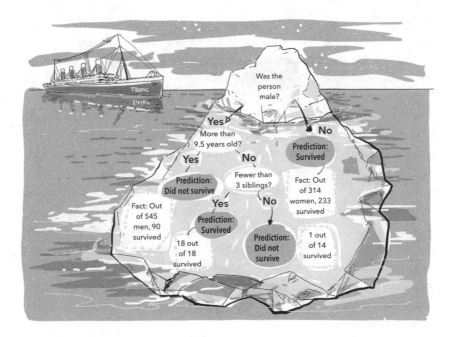

Figure 24
An algorithmically generated decision tree for predicting whether passengers from the *Titanic* survived the shipwreck. The image is based on a decision tree created by Wikimedia Commons user Stephen Milborrow, available at https://commons.wikimedia.org/w/index.php?curid=14143467.

distinctions are drawn are called *leaves*. Each person who goes through the decision tree based on the properties associated with them lands on one such leaf. The "prediction" they receive is determined by the fate of the majority of people on that same leaf. The tree, then, would predict that *all* women survived, even though that applies only to 74.2 percent.

Now after all that, how good is the tree's predictive ability? To find out, we again need a quality metric in addition to our test set that will determine how to aggregate the false and correct predictions into a single number. Before that, however, let's take a look at the number of false predictions the tree made for the set of training data. In this case, the simplest quality metric would count the number of people who received the correct prediction for each of the leaves on the trees. We can see that it was correctly predicted for 233 women and eighteen young men that they survived. It was also correctly predicted for 455 other male passengers and thirteen young men that they didn't survive. Out of the 891 passengers in the training set, then, the tree made 688 correct predictions, or 81 percent.

The training set is ultimately what the algorithm used as a point of reference to construct the tree, however, so it doesn't actually matter. If that were the only criterion, one could just as easily have divided the tree into greater and greater detail until each and every person received a correct prediction—resulting in a dreaded case of overfitting. The actual test of the tree's quality is how successfully the rules can be applied to the portion of data that *wasn't* made available to the algorithm. Let's run the 1,309 remaining passengers from the test set through the tree now to find out which leaf they land on, then compare the predictions to their actual fate.

The results: 339 out of 466 women total and twenty-four young male passengers were correctly predicted as having survived. Fifteen young men and an additional 664 out of 843 total male passengers that the decision tree assigned to the class of nonsurvivors did not in fact survive. Out of a total of 1,309 passengers in the test set, then, 1,042, or 80 percent, were correctly classified.

Should that number impress us? To decide, we run what is called a *baseline comparison*. You find the baseline, or line of comparison, by selecting the class that has the most observations, then use that class as the result for all predictions, meaning an accuracy rate of at least 50 percent. In our case, the largest share of passengers drowned. This means that if we had no other information about a person other than the fact that they were a passenger aboard the *Titanic*, we would have to assume they drowned. Applying this simple rule leads to a correct prediction in 61.6 percent of cases. The difference in our decision tree's 80 percent success rate is explained by the fact that other information was brought to bear.

Our decision tree did in fact find rules that also worked for the test set, or the portion of people that hadn't been considered yet. These rules speak to correlations, however, not causalities. A person didn't survive *because* she was a woman; if that were the case, *all* the women would have survived. Rather, the computer discovered that women stood a far *better chance* of surviving then men. Gender is correlated with survival but doesn't cause it. I'll come back to this point shortly.

In general, humans are more or less able to comprehend a decision tree of this sort, especially if things don't get too deep (or grow too tall!). In the preceding example, it is only the last rule about young men's siblings that remains somewhat puzzling. Why should it matter how many brothers or sisters someone was traveling with? The exact reason why will probably remain a mystery. Maybe it's because parents with fewer children were able to make it up to the deck more quickly. But it could also be pure coincidence: as I've explained, correlation is not necessarily causation.

## THE ETHICAL DIMENSION OF DATA SETS

Did you find it at all macabre, by the by, that I selected passenger data from the *Titanic* for my example? These are the real fates of individual people, after all—a fact that can go missing all too quickly when passengers are treated as data points. Reading through the Wikipedia synopsis of the shipwreck sends shivers down the spine. My goal with this book, however, is to show you what makes data scientists tick, and, as I mentioned before, the *Titanic* data set is frequently chosen for newcomers. Among other reasons, that's because there just so happens to be a small contest afoot regarding this particular data set, and as we've seen, nothing sets off computer or data scientists like a contest![6] You can find countless participants online, blogging or chatting about their method for arriving at the most precise predictions possible. Nearly every conversation about the data set, however, omits any mention of the passengers' individual fates. That may well be irksome, though I'd ask you to hold off passing judgment too quickly. Analyzing medication statistically proceeds by the same route, after all—that is, by deciding whether a new substance is more effective based on the overall data set, as opposed to individual fates. Yet while there are good reasons for doing so with medicine, you might still reasonably ask why you would do that for the *Titanic* data set.

Well, it's because we actually stand to learn a tremendous amount from this quite small and transparent set of data. Obviously, a set of clear-cut rules doesn't exist for telling us how each individual situation turned out, as with traffic violations. What we do know is that nearly every passenger who made it to a lifeboat survived and that there was actually preference given as to who should first receive a seat aboard the life rafts (which had been made available as prescribed by law, albeit in insufficient numbers). It's a preference that our algorithm also discovers: "Women and children first." Seen in terms of the data itself, this is exactly what we want machine learning to accomplish: to reveal hidden decision-making rules, even if they don't dictate the action 100 percent of the time and other properties may be involved.

In this case, the insights gained are historical in nature: they allow us to piece together a bit more clearly how things likely transpired at the time. On the other hand, there's the consideration that our analysis can hardly serve another purpose; its discoveries aren't applicable elsewhere. From an ethical perspective, it's also safe to assume that the data was probably not released directly by the passengers themselves. The data collection itself could be problematized, then, as could the fact that there exists no broader recognizable use for its analysis. I will continue to raise this and similar points in the following examples as they provide potent arguments for why, in the case of machine learning, as one of the mainstays of artificial intelligence, each and every citizen ought to be entitled to their say when it comes to decisions made about their person.

## FEATURE ENGINEERING: MANY TINY WIDGETS, ONE GREAT IMPACT

I've now introduced you to one possible example of a decision tree. In reality, computer scientists never stick with the first statistical model but are constantly trying to improve it. At all times, our guiding principle is the quality metric we've selected, which enters into a sort of competition with itself: Can I improve my own method any further?

One way of doing this is by *feature engineering*, a collective term for all the steps that go into deciding how exactly the input data is pieced together. Any number of participants in the *Titanic* contest, for example, improved his or her decision tree by searching the data set for affixes in passenger names as a possible indicator of nobility. Some added together the number of siblings and the number of parents, which were stored separately in the original data set, in order to come up with the total family size. Still others then went on to calculate the ticket price per person, as families paid a lump sum. All of this is accomplished manually, with very little that can be automated. Speaking about feature engineering in 2012, computer scientist Pedro Domingos described it as "often also one of the most interesting parts, where intuition, creativity and 'black art' are as important as the technical stuff."[7] Computer scientists rarely use just one method of machine learning. Instead, we work our way up from simpler methods, whose results are still relatively transparent and comprehensible, to more complicated methods. As with Knuth and Plass's text layout algorithm, we often don't have a general understanding of how these latter methods behave. We may be able to calculate what is going on with individual data points, but can no longer necessarily offer any insight as to what would happen with other, seemingly similar entries.

AT A GLANCE: DECISION TREES

Decision trees can be gleaned relatively quickly out of data. The algorithms behind them are in essence quite simple—though with so many decisions to make, they aren't trivial. As with the vast majority of the methods of machine learning, many of these decisions result in directives that technically are no longer algorithms but *heuristics*. While this means there's no way of knowing whether the decision tree we've found is the best one possible, such heuristics are still useful. The way in which a decision tree has been constructed hardly matters, after all, as its quality can always be described using test data sets and quality metrics. At the same time, it becomes clear that the quality of the test data must be as good as possible and that a meaningful quality metric must be chosen.

As soon as a decision tree has been trained, a very simple algorithm is tasked with making predictions about new data. The algorithm begins at the root of the tree, answering each question along the way and deciding in each case whether it should carry on. Upon reaching a leaf, it gives a prediction that corresponds to how the majority of the people in that group behaved. Alternatively, it can state the share of people in a given class (in the case of the *Titanic*, the ratio of survivors to nonsurvivors).

Why have I bothered going into such detail with all this? It is because the example clearly illustrates each of the individual steps involved in developing algorithmic decision-making systems, or what I've called the *long chain of responsibility*. The following observations are important, as they can be generalized:

1. Machine learning is a tool. At no point does some form of "magic" come into play that generates truth directly from data. The method used to construct a decision tree follows a detailed set of instructions, step by step.
2. A number of other decisions have to be made before a decision tree can be constructed. How large will the training and test sets be, respectively?

At what point has there been enough differentiation? It's only by experimenting with these sorts of decisions, a process referred to as *hyperparameter tuning*, that significantly better decision trees can be developed. This is important to realize because a great deal is attempted here manually, so to speak, so that the statistical model will function smoothly afterward. It represents another juncture at which the algorithm's design can be influenced.

3. There is actually both an algorithm and a heuristic at work with decision trees: a heuristic for building it (which in the media is often actually called an algorithm) and an algorithm that then goes through the decision tree with the new data to arrive at a decision.

4. Both the heuristic and the algorithm are simple enough that neither requires an "algorithmic safety administration." The methods' many variations were programmed ages ago and have been used by hundreds of thousands of people, presumably without a hitch. Rather, it is the overall process of modeling and all the decisions such a process entails that might have to be checked, depending on the circumstances.

There are also important observations to be made about selecting data. First, as with many other data sets, the information about passengers aboard the *Titanic* was incomplete and may have been wrong in some cases. Second, the data contained a form of discrimination: women and children were given preference when filling the life rafts. If one were to use the tree shown earlier to make decisions about who should be saved the next time disaster struck, *that discrimination would only grow in magnitude*! If in each case the decision tree were to opt for the largest group on the leaf, it's possible that in the end only women, girls, and young boys would wind up on the rafts. Depending on how the statistical model is used, then, *a set of data containing a form of discrimination will project that discrimination forward into the future, and even strengthen it*.

In terms of the overall method, we can note that data scientists do not construct a single statistical model once and for all. Rather, the initial model is often continually being refined, whether by adjusting the initial selection of data through feature engineering or opting for different methods. In the process, *the quality metric becomes a sort of gold standard*, deciding when the designer is satisfied with the software and no longer needs to improve it.

And why, you may ask, does it matter for you to know just how many manual—read, human-made—decisions go into machine learning, and how many buttons and knobs there are to fiddle with in the process? Once again, we're back with MacGyver, tinkering away.

It matters because it means that we aren't all at the collective mercy of a machine's verdicts: computers are not simply applying some pure form of mathematics that is able to identify facts objectively and indisputably. *It matters too because it means that any number of decisions might be wrong and that even without deep technical knowledge, you both can and ought to have a say in some of these questions.* That is why I'm now going to have you, at least this once, make this kind of decision. And to do that, I am turning you into a machine—a support vector machine should do nicely.

HOW TO BECOME A MACHINE

Have you ever been unlucky enough to have to sit through a job interview that had no point whatsoever? It hardly matters which side of the table you're on. I for one can certainly recall the hours on end spent in my student co-op, yakking with a potential housemate who nobody liked anyway for "just another fifteen minutes" about a highly innovative (and perhaps just a teensy-weensy bit outlandish) concept for refinancing a washing machine.[8]

Wouldn't it be wonderful if we were able to discover which properties have led to successful job placements or applications simply based on applications from years past? Better yet, what if machines could make those decisions independently of the person's reputation or social standing—that is, without our having to worry about discrimination? Just imagine it, a world of purely objective decision-making—and far fewer pointless interviews.[9]

It must be our lucky day, because some cool new data has just come in about the latest pool of applicants to the imaginary company Good Work. Of the twenty-seven people that applied, thirteen were hired and remained at their position for at least two years, which the company manager has set as the deciding criteria for a successful hire. The two most important factors contributing to an applicant's success have also been identified: their number of years of experience and the number of years they've spent unemployed. The properties appear as the axes on the chart in figure 25.

As you can see, in the past, people with long periods of unemployment tended to be less successful in their applications or subsequent careers at Good Work (as indicated by the dark square frowny figures).

It's now up to you to train a support vector machine based on the data. A what, you say? And just how exactly am I supposed to do that? No programming skills? No problem. Take out a pen and ruler, then draw a straight line across the chart that divides the unsuccessful (square) applicants from the successful (circular) applicants as neatly as possible. The line can run any way you like: diagonally from top left to bottom right, right to left, straight down from top to bottom, whatever. The only requirement is that it is *straight*. Ready, set, draw! (Cue *Jeopardy* music.)

Well? Where does your line run? Whatever you decided, my secret telepathic powers as a professor tell me that you weren't able to find a perfect solution. That may also be because I selected the data set so as to make it impossible; there will always be unsuccessful applicants on the side of successful applicants, and vice versa.

The smallest possible number of errors of this sort is four. If it doesn't matter which type of error occurs, there are at least three such optimal lines that achieve the result.

Figure 25
Draw a straight line that divides the dark square frowny figures from the bright
circular smiley figures as neatly as possible.

If, dear reader, you did participate in this little test and drew a line, then
in effect what you've done is train a *support vector machine*.[10] The name is
complicated, but what it does is not. The underlying heuristic searches
out a line that divides the two applicant groups as best as possible, leaving
the maximum number of objects (data points) from one group on one side
and the maximum number of objects from the other group on the opposite side. This part of the process is easily automated and performed by
the computer. In the real world, what happens is usually somewhat more
involved, as it is often more than two properties that are deemed important.
In mathematical terms, we are searching for a (hyper)plane that divides the
two groups from each other as well as possible; you can picture it like a
knife cutting through the data. This is what represents the statistical model,
regardless of what the dividing line or plane looks like. With it, any new
data point (or applicant) can easily be categorized.[11] If the new data point

falls on the side of the successful applicants, we can assume that the algorithm will predict the applicant's success. If it falls on the other side . . . well, we can probably spare ourselves the chat.

Figure 26 shows the lines students in my Introduction to Socioinformatics lecture course drew during winter semester 2018–2019, the smaller portion of which were optimal in the sense discussed here.

Only lines B and F result in the minimum number of errors, at least as twelve of my students drew them. What I find fascinating, however, are Lines I and G, which are clearly intended to make sure that only successful candidates receive an interview. In each case, the student who drew the line was forced to accept that seven people who would in fact have made for successful employees would be falsely categorized.

And now for the crucial question: You receive a new application from a person with five-and-a-half years of professional experience and a total of four years unemployed. Do you invite them in or not?

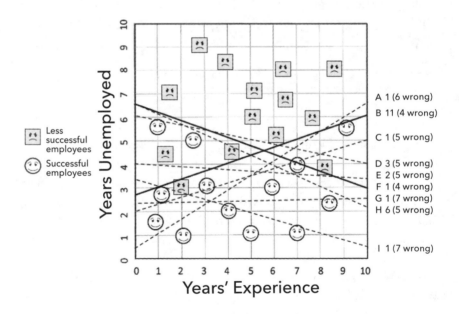

Figure 26
The dividing lines drawn by my students. Each line is identified with a letter followed by the number of students that selected it. The number of data points lying on the "wrong" side is in parentheses. The solid lines in the figure are optimal, with only four incorrect classifications.

Figure 27 indicates with an X where the new data point, or applicant, would appear on the chart. Compare its position to the line that you drew before and make your decision: Do you ask the applicant in for an interview? Figure 28 shows what my students decided.

Six students drew their lines so that the new data point fell on the side of the unsuccessful square frowny figures. In figure 27, those are Lines A (one student), C (one), E (two), G (one), and I (one). The other twenty-one students have the new data point on the successful side with the other circular smileys. In other words, 22 percent of students would not have invited the new applicant in for an interview.

The students in my lectures aren't always so dismissive; generally it's around 5 percent. What's truly interesting, however, is how differently the lines were drawn! Why is that?

Figure 27
The "X" marks the spot where the new data point falls on our chart.

First, it has to do with the wishy-washy nature of the task itself. Unfortunately, more often than not data scientists find themselves tasked with vague jobs that haven't specified what exactly is supposed to be optimized. Once again, in order to determine what a "good" decision entails, it first has to be operationalized, or made measurable. Data scientists have developed an entire series of quality metrics to this end because it isn't really something machines can take charge of. *We as people have to define what a good decision entails for the results of machine learning to be able to generate good decisions in the first place.*

The quality metric—alongside something else called a *fairness metric* that I'll come back to later—are fundamentally what determine what it is the computer learns. And happily for us, very little technical know-how is needed here to have an opinion! That's because, in computer science speak, it is about operationalizing the *social concept of a decision's quality.* Sound boring? Au contraire!

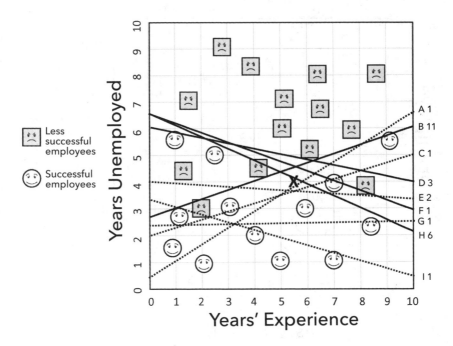

Figure 28
The dotted lines have the new data point with five-and-a-half years of experience and four years of unemployment fall on the unsuccessful side, while the solid lines set it on the successful side.

AWKWARD INTELLIGENCE

As chief decision scientist at Google, Cassie Kozyrkov certainly has the hippest job title of any woman I've ever heard of. It's music to my ears to hear her talk about how it is often the wrong people (data scientists) who are entrusted with the question of what exactly AI should be able to do. Meanwhile, the ones who should actually be making the decisions scamper off, mumbling something like: "Go sprinkle machine learning over the top of our business so . . . so good things happen."[12] In all seriousness, though—nothing will come of it.

In her article "The First Step in AI Might Surprise You" (that's right, folks, she also does clickbait!), Kozyrkov likens it to dog training. You may not have to know how any given training method will help the dog learn, but you do have to know whether the goal is to train a police dog or a

sheepdog, for example. And if the dog is supposed to look after sheep, it
might be a good idea to check whether you actually have enough sheep to
teach him what they are in the first place.

So how do we tell computers what they're supposed to learn? One thing,
of course, is for the training data to include information about when whichever property is to be learned was observed. Yet in the vast majority of
cases, no algorithm on planet Earth will manage to come up with a decision
rule that is always valid and clarifies beyond all shadow of a doubt which
measurable properties will lead to others. That's especially the case when
it comes to predicting how individuals will behave in the future, which as
a rule depends on the person under consideration but also on the situation
in which she finds herself. If it isn't possible to find a rule that applies 100
percent of the time, however, it means the various errors a decision may
result in have to be weighed against each other.

Is it just as bad for a person to mistakenly receive a job interview, for
example, as it is for them to mistakenly *not* receive the interview? Communicating with the machine in order to carry out this type of weighting
occurs via the quality metric. Whenever a data scientist believes she's found
a statistical model with good decision rules, the quality of those decisions is
measured with a test data set aided by the quality metric. If the test results
prove satisfactory, training can stop. If not, it's back to all the various levers
and pulleys data scientists use to make the relevant adjustments—or maybe
even one of the many other methods machine learning has to offer.

The quality metric thus plays a deciding role in training. To continue
with our analogy, it is what decides when the dog gets a treat. A police dog
latches on to the cuff of a thief's pants to prevent him from running out of
the store? Good dog! A sheepdog sinks his teeth into a sheep's leg and drags
it over to the herd? Bad dog!

In this context, Twitter user Custard Smingleigh (@smingleigh) shared
a charming anecdote that explains quite well how a quality metric can help
a system optimize itself on the spot.[13] It all began when he went to improve
Fenton, a robotic vacuum cleaner of his own design that was moving too
slowly. As you might expect, Smingleigh also wanted to protect his furni-
ture, so he sketched a quality metric that rated speed positively and collisions, registered via sensors, negatively. The results were fantastic: Fenton
simply buzzed and whirred its way around the apartment backward, crash-
ing into everything except the cats, who sprang out of the way quickly

enough. Why? Because Fenton didn't have any sensors in the back (no hind-sight there!). The training rewarded high speeds that didn't lead to *registered* collisions, so from then on Fenton increasingly traveled in reverse, with no sensors to detect a collision of any kind. Problem solved! Then again . . .

Smingleigh used a neural network with Fenton, by the way, a topic I'd like briefly to turn to now. In fact, it is neural networks we have to thank for all the conversations we're having at the moment. With their help, prob-lems whose answers long proved elusive are now within reach—image rec-ognition, for example, or machine translations that are actually worth their mettle. In what follows I'd like to show you how even with neural net-works, there is no automatic mechanism at play that simply knows where the "truth" lies in the data without any human assistance. These methods also depend on how data scientists present them with information: the data we select and which problem we actually want to be solved optimally by which yardstick.

## A BRIEF ASIDE ON NEURAL NETWORKS

In his tweet, Smingleigh doesn't talk about a quality metric but a reward function. "How . . . do you reward a Roomba?" one user asks in the thread.

"They 'love' vacuuming up Cornflakes," Smingleigh jokes at first, before going on to explain in detail. As with other terms we've discussed, *neural network* is a big name for what is in truth a relatively simple mathematical structure. In this case, the structure is composed of mathematical functions arranged in rows. The first row of functions is fed input data—in this case, the current speed of the Roomba, whether the bumper sensor is active, and the camera image. The camera image may be preprocessed to show where exactly potential obstacles lie and how far away the robot is from them.

Once these input data have been processed by the first row of mathematical functions, the results are then entered as input data into the second row, whose results are then entered into the third row, and so on. The final row corresponds to actions the Roomba might perform: stop, turn right or left, accelerate forward or backward, and so on.

In principle, the mathematical functions from the first row can use any input data provided, but each input is weighted differently. Some weigh the camera image more heavily, while others evaluate the bumper sensors

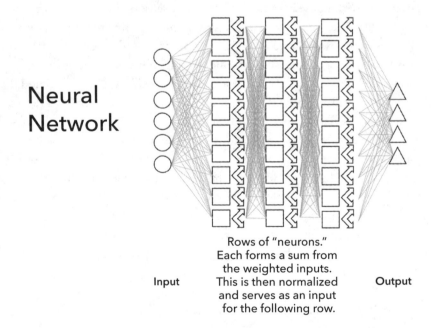

**Neural Network**

Rows of "neurons."
Each forms a sum from
the weighted inputs.
**Input**        This is then normalized        **Output**
and serves as an input
for the following row.

Figure 29
A simplified diagram of a neural network.

or some other combination that has been selected. The result is then run through a second function, which brings the weighted calculation back to somewhere between 0 and 1, a process called *normalization*. Every combined set of calculations and normalizations forms a "neuron," based on the idea that our own nerve cells receive different sensory inputs and then either "fire" (pass on a signal) or do not—thus the normalization to values between 1 (fire) and 0 (remain inactive).

In addition to being linked to sensory cells, human nerve cells can also be linked to each other. In neural networks, this is represented by making the output of the previous row the input of the next. Between the rows, too, the inputs are again weighted, with the mathematical functions paying greater attention to some inputs than others. The final row presents the possible actions the Roomba might take. Actions with values close to 1 are executed, then rated according to the reward function. If the situation and the selected action allow the Roomba to move around quickly with zero collisions detected, the cells that "voted" for the action (delivered a value closer to 1, thereby contributing to the action being selected) are connected more tightly to the other cells that contributed to the action. In other words, the weighting shifts for each of the inputs that made a contribution—a form of positive feedback for the nerve cells that made the "right" decision. Just how much the weighting is changed, and at what moment—therein lies the art of the data scientist, who busies herself tweaking knobs, adjusting scales, or flipping the switches of the neural network.

If the action leads to slower speeds or a registered collision, on the other hand, the weighting that led to it is weakened (negative feedback). The "nerve cell connections" that did it "right" get a treat; the others get a spank. Phew! That was a fairly weird analogy; I hope it sticks.

### WELL, AREN'T YOU ACCURATE!

Let's come back now to the problem of choosing the "right" quality metric for our support vector machine so as to identify the best candidates for a job interview. Before, I said we could simply add up the total number of correct decisions that resulted from the line drawn for applicants in the test set. That includes decisions to invite applicants who proved successful, as well as decisions *not* to invite those who proved unsuccessful. Remember—the test data set is made up of individuals for whom the company knows already whether they were a successful hire or not. It's only the algorithm that doesn't know, and makes its decision based exclusively on other properties it does know about the applicants.

The sum total of all correct decisions tells us what a system's *accuracy* is. Now if you hear someone say, "My AI decides with an 83 percent accuracy rate," you will know that 83 percent of all decisions made were correct. When you hear talk of accuracy, you also know that a quality metric is involved, which measures the prediction's relative proximity to the truth.

The thing about accuracy is that while it may seem quite straightforward, it can be quite tricky to interpret. It's easy to demonstrate very impressive levels of accuracy, especially when one of the two groups contains many more data points (i.e., applicants). Many companies, for instance, will receive a thousand applications for a single position, meaning a success rate of one in one thousand. It also means that if a prediction system says simply not to invite anymore, it will make the right decision 999 out of 1,000 times, for an accuracy rate of 99.9 percent! This particular situation and many others explain why there are around *twenty-five different quality metrics* in total. Twenty-five! And as we've just discussed, choosing the right one depends on exactly what kind of situation it will be used in.

This is also what makes it so difficult to find pretrained software systems for predicting human behavior on the market. It is downright improbable that the quality metric with which a given system was trained will correspond to your company's situation. In the end, you get a dog that's been half-trained as a police dog and is now suddenly supposed to herd sheep. Nor may that be something you notice, depending on the circumstances. You're the only one with a knowledge of the exact conditions under which the system will be used, which means you have to chime in on the process; otherwise, it's left to data scientists to decide for themselves which quality metric will obtain what are ostensibly the best results.

Even more important, however, is the fact that the quality metric also makes moral judgments without any sense of the actual context, as the next example shows.

### HOW ETHICS ENTERS THE MAINFRAME

Remember how my coworker and I had sat there dumbfounded at the results of a recidivism prediction software used in US courtrooms? How would that look if we trained a support vector machine? What criteria would you use to optimize it? In figure 30, I've used the exact same data points as in figure 25 but given the smiley/frowny icons different meanings. Imagine that the presence of two different hormones in the blood allows us to predict to a certain extent whether or not someone will commit a crime—we'll call one *criminoleum* and the other *peacetrogen*. The bright square icons stand for criminals, the dark circular icons represent law-abiding citizens.

Now draw another dividing line between the data points, which in the following step will be used to classify other people. Is the line different from the one you drew before for the job applicants? Did the context change anything about what you wanted to optimize for?

This situation confronts us with what turns out to be a difficult balancing act between opposing interests and values. On the one hand, society has

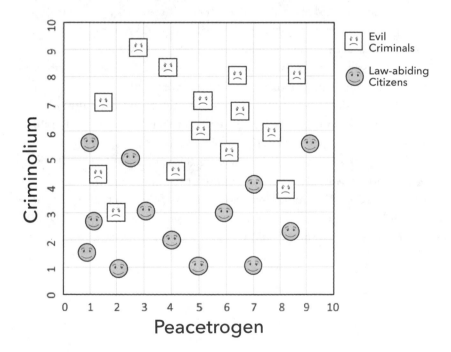

Figure 30
A fictional data set for criminals and law-abiding citizens reflecting two properties: the fictional hormones criminoleum and peacetrogen.

a justified interest in identifying as many criminals as possible. At the same time, it's clear the innocent should be protected. All the way back in 1760, English jurist William Blackstone coined the maxim, "It is better that ten guilty persons escape than that one innocent suffer."[14] These days, we have former US vice president Dick Cheney, who, when interviewed by Chuck Todd in 2014 about the Senate torture report on conditions at Guantanamo Bay and other camps, gave the following opinion: "I'm more concerned with bad guys who got out and released than I am with a few that, in fact, were innocent." When Todd persisted, asking whether that still applied when an estimated 25 percent of those held were innocent, Cheney replied, "I have no problem as long as we achieve our objective."[15]

The optimization function's moral underpinnings, it seems, would lead Blackstone and Cheney to make very different decisions. Blackstone's line would ensure that no one who was innocent ended up on the wrong side, while Cheney would draw his so that every last criminal was caught.[16]

The example demonstrates that choosing a quality metric *always* entails a *moral consideration*—namely, which wrong decision weighs more heavily. Nor is it possible simply to *abstain* from choosing. Our discussion of accuracy, for example, implied that both mistakes were equally bad: the number of correct decisions was counted up without taking into account whether someone had been correctly labeled as a criminal or an innocent civilian. Both correct decisions were ascribed the same importance. That can change when weighting is introduced, as with Blackstone and Cheney. But who exactly should be in charge of determining this weighting, and how should that be decided? This means yet another cog in the machine, a new scale that will always be of an ethical nature, never "neutral" or "objective." The question now is this: Would you weigh things like Blackstone? Or like Cheney? The choice is yours!

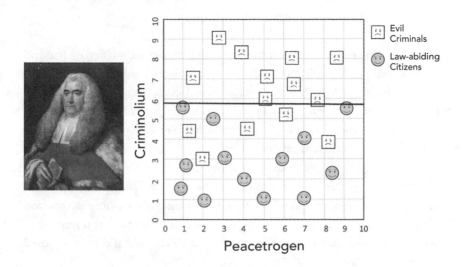

Figure 31
Legal scholar William Blackstone would have drawn his line so that every innocent civilian ended up on the "right" side, even if in this case it means that five people who actually belonged behind bars went free.

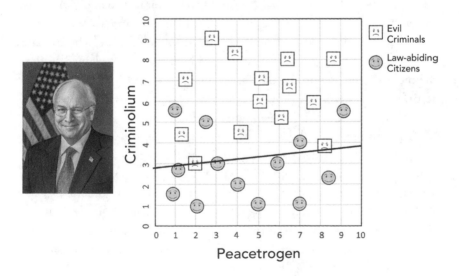

Figure 32
Dick Cheney, on the other hand, finds it much more important to apprehend criminals than to protect the innocent. His line looks different: he catches all the criminals, but also puts eight innocent people in jail.

THE SOCIAL CONTEXT OF QUALITY METRICS

Now that we've identified accuracy as a first important quality metric and familiarized ourselves with the difficulty of making moral judgements, I can sketch for you how it is that a handful of US states could use recidivism prediction software that seems so ill-suited to the purpose.

In 1998, software was developed for rating how likely criminals were to commit another crime after their release. This rating served in turn as the basis for decisions about providing different kinds of assistance. A questionnaire was created for use as input in the software, alongside data banks with information about previous crimes. An algorithm whose design was kept secret learned a set of decision rules (based on a set of training data), which were then used to assess the level of risk a new set of criminals posed of committing crimes in the future.

In our fictional job application example, people were sorted directly into one of two categories: successful or unsuccessful. Here the situation is somewhat different: people were first sorted according to their risk value, then, in a second step, two thresholds were established. Individuals whose risk value stayed below the first threshold were assigned to a low-risk category. Those whose risk value lay between the first and second thresholds were placed in the medium-risk category, and anyone whose value exceeded the second threshold was labeled high risk. As explained earlier, the method itself resembles how insurance plans work. People are rated according to the risk posed by others in the same class—that is, the actual rates of recidivism that have been observed in the past.

The company allowed the software to be tested in studies, reporting that it resulted in correct decisions 70 percent of the time.[17] "Aha, accuracy," you must be thinking to yourself. "I know about that, I get it." In this case, accuracy wasn't the quality metric, however. Instead it was something with the bombastic title *area under the receiver-operator characteristic curve* (ROC AUC). Don't let yourself be intimidated; ROC AUC is just a percentage. To determine the percentage, you consider people from the test data set *in pairs*, in each of which one commits another offense and the other does not. Data scientists are able to form these pairs because the test set is composed exclusively of data points whose results have been observed: we know who committed another crime and who did not. We don't tell the algorithm that, however.

We then test for each pair how often the algorithmic decision-making system manages to assign the person who did commit another crime a higher risk value than the other. That was the rate that lay at around 70 percent for the COMPAS algorithm. The rate meant conversely that for around 30 percent of the pairs, the person who managed to regain their footing in society received a worse rating than the person who went on to commit another crime. Strikingly, that number was about the same for a second, quite different risk score: in addition to letting you predict the *general rate of recidivism*, the software also lets you assess risk for *violent crime*. That meant there were two different scores and two different sortings, both with a value of around 70 percent in the ROC AUC.

Personally, I don't find this percentage all that great in and of itself. If you were to assess the same group of people at random, you'd stand a 50 percent chance of making the right decision, which makes 70 percent nothing to write home about.[18] Still, 70 percent might be tolerable for the situation in which the software was originally used. If five people are released from prison on the same day but there's only one social worker available to help with resocialization, an algorithm that draws the right conclusions for pairs of people stands a decent chance of finding the person most in need of assistance. In this particular social situation, what matters is that we are forced to decide: now, in this moment, a social worker has to be assigned the person most in need of help.

Today, however, the very same risk assessment system is used for *pretrial classification*. That's important because pretrial classification works differently as a social process: the goal here is to keep people who judges assume will commit another offense—that is, those in the highest risk category— behind bars. Can we infer from this that an algorithm with a ROC AUC of 70 percent will also sort 70 percent of recidivists into the high-risk category? No, no, and, once again, no!

Let's imagine for a moment that all of the people who have been rated by the algorithm are arranged according to their risk value—no-goodniks at left, those in the middle at center, harmless souls on the right.

The trouble is that with the ROC AUC, the model was trained to sort pairs in the middle of the spectrum correctly, as well as those at either end. Any pair that was sorted correctly received a reward or "treat" regardless of where it fell. That means that if the algorithm had to choose a weighting for a given property that either moved two more recidivists into the high-risk group or positioned five recidivists in the middle further left, it would choose the latter option.

For readers who want to know exactly how this works, the following illustrations show what the ROC AUC is actually measuring.

An algorithmic decision-making system has assigned criminals a risk value for recidivism and sorted them accordingly. Those with a high-risk value are on the left. The color and shape indicate whether or not the person commits another offense: bright circular smiley icons stand for those who do not, the grey box icons for those who do.

When the algorithm assigns a higher risk to someone who is a recidivist, it's a "good pair"; if not, it's a "bad pair."

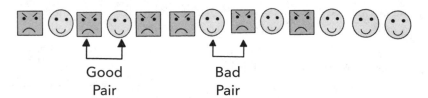

The ROC AUC states the share of good pairs out of all possible pairings. There are six recidivists and seven nonrecidivists, making for forty-two pairs total. The figure at the far left (a recidivist) stands to the left of seven circular icons, or nonrecidivists, making for seven good pairs. The second square icon from the left creates six good pairs, as it stands to the left of six circular icons. The next two grey smiley icons make five good pairs because they have received a higher risk score than five circular smiley icons. The last two add only four and three good pairs, respectively. Out of a total of forty-two total pairs, then, thirty are good, or 71 percent. That is the ROC AUC for this sorting.

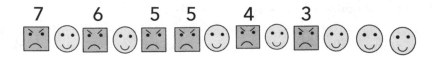

But now, as with most social processes and especially in court, it is only the people at the far left that interest us, a group that wasn't given any special attention during training. As the group that is likely to be incarcerated, what matters here is the rate at which this group actually recidivates. This percentage is called the *positive predictive value*. As we've discussed, *positive* is used here as it is with medical tests: we are looking specifically at those who do recidivate, not indicating that their behavior is somehow positive.

When predicting crimes of a general nature, about 70 percent of those in the high-risk group went on to commit another crime; in this case, the percentage value of the ROC AUC and the positive predictive value coincide. When it came to predicting violent offenses, however, a mere 25 percent of those in the high-risk group actually became recidivists. In other words, three out of four people from the test set who the algorithm assigned to the high-risk group *didn't* commit another offense! I have these numbers from a study the company conducted itself, by the way, which means it is entirely aware of the fact that classifying a person in the high-risk category does not necessarily mean they actually represent a higher risk, in the sense of "will almost certainly become a recidivist."[19]

But why then was the ROC AUC selected as a quality metric? In the end, that can't be explained by someone on the outside. A couple of potential reasons do come to mind, however. First, within the computer science community, the ROC AUC is often considered the most valuable quality metric out there. Second, it is in fact an appropriate metric for some social processes. It was originally developed as a way of quickly identifying the most at-risk person within a small, randomly assembled group of people, for example. If the algorithmic decision-making system is intended for something like pretrial classification, however, where only the high-risk group is relevant, it would be better to use a quality metric that includes how many people in a particular category do actually belong to the group of recidivists. If the quality metric doesn't match the situation, the value can be as high as you like and it still won't matter. Think back to the analogy of dog training: an algorithm that was trained with the wrong quality metric is like telling a police dog to herd sheep. Once again, we've found the OMA principle at work. The *model* of the world must be attuned to successful *operationalization* and the *algorithm* itself.

A third possible reason for using the ROC AUC as a quality metric is mundane: It's usually quite easy to achieve high percentages with the ROC AUC, without that necessarily meaning that quality is just as high in every regard.

Whatever the case may be, the example is a striking demonstration of the fact that especially when the state purchases algorithmic decision-making systems, it urgently needs experts by its side who are paying attention to these aspects.

WARNING, TERRORIST!

Nor is COMPAS the only instance of a decision-making system being described in terms of quality metrics that make it difficult to grasp intuitively the extent of the wrong decisions. The same dynamic was at play with SKYNET, a system for identifying suspected terrorists whose existence was first brought to the world's attention by Edward Snowden.[20] A document circulated online—a series of presentation slides—shows how methods of machine learning were used to assess the risk of terrorist behavior for fifty-five million people based on their cell phone data.[21] Strictly speaking, the system identified couriers that went between terrorist groups, an activity for which cell phone data was ideal as it showed when a person was active, where they were moving, who they were in touch with, and how they communicated with each other. Swapping SIM cards on multiple occasions, regular travel, communicating with groups that didn't talk much among themselves otherwise, nighttime activity: all were possible signs a courier was involved.

To learn the properties of these couriers, a training data set was needed that included enough couriers and noncouriers. The initial courier class, however, contained just seven lawfully convicted terrorist couriers. Seven. Out of fifty-five million! Predictably, it was an impossible task; the algorithms of machine learning have a tremendous appetite for data.

Neural networks in particular require thousands of data points for both classes. With SKYNET, this meant that decision trees were used at first, but they also require much more than a handful of data points for the category that is to be predicted. The critical issue was raised in the leaked document

and led to a second step in which "selectors" were used in addition to law-fully convicted terrorists to train the algorithm. The slides don't go into any greater detail about who exactly counts as a selector; presumably, it's either known suspects or people in direct contact with those already convicted.

The statistical model that resulted from the new data then assigned each of the fifty-five million cell phone users a risk value. The system used a binary classifier—one that gives a thumbs up or thumbs down—which meant that a threshold also had to be defined. The threshold was set arbitrarily so that 50 percent of the "terrorists" (those convicted + selectors) lay to the left of the threshold and 50 percent to the right. This is a bad decision because even out of the known "terrorists" it means that only 50 percent would have been identified with the threshold. To make matters even more complex, don't forget that all these annoying quotation marks are meant to recall the fact that some might not even be terrorists. Choosing such a threshold would be justified if—and only if—it decreases the false positive rate considerably—that is, the rate of people wrongly accused by the algo-rithm. As it turns out, we do know how high that value is, according to the source: 0.008 percent.

Sounds good, right? Great work, what a small percentage!

But wait a second. How many people were there in the overall data set again? Fifty-five million—almost none of whom were actually convicted of being a terrorist. And how many people is 0.008 percent? It's 4,400 $(55,000,000 \times 0.008 = 4,400)$! "Yeah, yeah," a true believer in data would reply, "but they might also all be terrorists!" Before you nod your head in agreement, though, I'd like to introduce you to the suspect with the highest risk value out of the entire group, Ahmed Zaidan. The slides list him as a member of Al-Qaeda and the Muslim Brotherhood, which according to everything we know today is wrong. In reality, Ahmed Zaidan is a journalist for the news channel Al Jazeera. Writing about the algorithm's results, Zaidan explains how his work as the bureau chief for Al Jazeera's branch in Islamabad made him the lead suspect: he was in fact moving regularly through suspicious areas and did interview bin Laden and other terrorists. In his role as a journalist, he saw himself as a mediator, "especially when there is a meltdown in communication, and conflicting parties are resorting to everything but dialogue to resolve their differences."[22] He writes that the algorithm's results "put my life in clear and immediate danger," and voices concern for the safety of journalists the world over who carry out the same mediating role.

What's especially important to recognize here is that as before, establishing the threshold at which a person becomes a suspect represents a moral decision. *This isn't a decision some algorithm can make on its own*, nor can a group of data scientists, for that matter. Rather, the decision has to do with how we as a society judge the two possible wrong decisions relative to one another: an overlooked terrorist versus an innocent civilian. I don't know what you yourself would say, whether you'd find it appropriate in light of the danger of further attacks to assign 4,400 people to the high-risk category—even with the consideration that innocents may come under suspicion—or whether you would weigh things differently. I do view it as imperative, however, that such a decision be carried out using democratic means, by a qualified body that is given full insight into the underlying data, the algorithms, and their results.

These last two examples present clear and unmistakable cases of quality metrics that involve fundamental ethical positions. Moreover, it's evident that it is not primarily technical know-how that matters but rather *society* defining for itself what a good decision entails. That's one decision we shouldn't let anyone take from us: algorithms have no sense of tact and no sense at all of proportionality. All they have is their quality metric, and as we've just seen, this too depends on the social context in which the results will eventually be used.

## THE ETHICS OF THE GROUND TRUTH

There is one other point I'd like to discuss in this context, one which may already have occurred to you. It concerns the question of the underlying data and the ground truth.

I've almost certainly used the terms *recidivist* or *recidivism* a dozen times or more in the preceding paragraphs. How do you measure recidivism, though? One does so by defining a certain period of time (usually two years) following a person's release from court or jail. If he or she commits another crime during that window, it counts as recidivism.

There are two problems with defining the ground truth in this way. First is the fact that committing a crime in and of itself isn't enough for someone to be labeled a recidivist; the person must also be caught, accused, and sentenced for the crime. Especially in the US, however, certain groups of people are stopped and searched much more frequently by the police, and there generally exist certain kinds of crimes that are easier to discover.[23] Nor are conviction rates equal among the country's different population groups. It was for this very reason that the American Civil Liberties Union (ACLU) made an appeal in 2011 for algorithms to help with different phases of the sentencing process.[24]

Things wouldn't be so bad if, as is possible with product recommendations, the systems started out at a lower level of quality and were then able to improve over time. When users click on a recommended product or ideally leave a little feedback if they purchase it, they are signaling whether or not the product interests them. The millions upon millions of responses the underlying systems receive allow those systems to improve. The undue amount of trust that the state and private industry place in the predictability of human behavior likely comes from these sorts of applications, which have in fact proved themselves quite helpful—in a completely different context.

Yet one serious problem with predicting human behavior both in general and in particular (e.g., criminal behavior) is that in most cases, it is not all that simple to use the observed behavior as feedback. This is because *feedback* is quite often *one-sided* in nature. Take recidivism: If the machine classifies somebody as low risk and the person is free to walk, one can observe whether he or she recidivates. If they do, the machine's algorithm can be changed to prevent that from occurring the next time. Things become more difficult, however, if someone is wrongly sorted into the high-risk category and receives a longer prison sentence as a result. After serving their sentence, they are a different person than before; they've now missed more years of work, while their criminal record further diminishes their chances of landing a job. This may in turn increase the likelihood of another offense—which might seem like a confirmation of the prediction, but in truth represents more of a *self-fulfilling prophecy*. By the same token, if the person doesn't commit another crime after he or she is released, it's not necessarily a sign that the algorithm made a mistake. It is possible that the prison sentence led the person to change their life, after all. In any event, it becomes clear that for those in the high-risk category, no straightforward corrective mechanism exists for comparing the algorithm's prediction to a person's behavior because the prediction and resulting action might alter their behavior in the future. This means there is a constant risk of creating self-fulfilling prophecies, or a prediction having the exact opposite effect. In short, one-sided feedback means that if an algorithm makes a wrong prediction, it can learn and be adapted in only some of the many possible scenarios.

Is it bad when such one-sided feedback is the only kind that exists? To answer this question, I'd like to return to the wily little robot Fenton. Fenton's optimizing function did gradually improve, although as soon as it began to travel in reverse, it unfortunately received only one kind of feedback, and that was *Faster, Harder, Scooter!*[25] It learned nothing at all about the cases where it predicted the path ahead was clear but the leg of Grandma's dressing table actually stood in the way: How can hindsight be twenty-twenty if you literally don't have any? Anyone who's witnessed children who only ever receive praise or criticism will know they rarely have the most even-keeled of personalities.

One-sided feedback thus makes for a weak foundation when training algorithmic decision-making systems, especially when there is relatively little training data available at the outset. Yet when it comes to predicting risk or success in humans, *one-sided feedback is more the rule than the exception*. Those using such decision-making systems will often try to avoid people who pose greater risk, after all, and seek out those with a high potential for success. That leaves job applicants who supposedly show little promise unable to prove that they might in fact have done well in the position; those whom a machine has predicted to be a greater credit risk unable to demonstrate that they could have paid off their credit card; and children with "low" learning potential unable to demonstrate that they could in fact have succeeded.

So what is machine learning good for, then? The cases discussed so far—terrorist threats, creditworthiness, and selecting job applicants—don't exactly seem like prime examples. Yet there are areas where machine learning works quite well, and such systems have been shown to work better than people. In chapter 6, I introduce you to two.

## MACHINE LEARNING VERSUS PEOPLE

As it is, dear reader, each of us probably uses multiple products that rely on machine learning every day, most often with perfectly respectable results. For one thing there are all the different sorts of recommendation systems out there, whether for products, advertisements, search engine results, or personalized news feeds on social media. The recommendations they come up with usually fit and sometimes are even really good, but "good enough" at least to serve as ersatz human salespeople, ad agents, and librarians. In what follows, I discuss two further examples in which machines are demonstrably better than people—image recognition, and as a special case detecting skin cancer—to then draw conclusions about the general conditions under which machines hold an advantage.

### IMAGE RECOGNITION, OR WHAT AM I ACTUALLY SEEING?

To get a better idea of what's going on with image recognition, think back to the last time you visited a zoo, ideally one with marine life. I guarantee you overheard a kid turn to his poor parents at some point while watching a group of brown-haired animals splash about playfully and ask, "Mommy, what is it? Is that a walrus?" "No, that's a seal," comes the reply. "Or wait, is it actually a sea lion?" Now an older woman butts in, "No, no, it's a leopard seal, you can tell by the nose!" Turning to the sign in front of the enclosure, you read that you are looking at none other than a fur seal. Supposedly it's possible to learn how to tell the critters apart.[1] I myself fail every time and am now just waiting until my two kids are finally able to read the signs themselves. If you're able to distinguish between them all just fine, thank you very much, because you happen to hold a doctorate in veterinary sciences in the field of Pinnipedia, maybe try your luck hunting down a couple pounds of edible fungi next fall with a book on mushroom identification.

Assigning the things we see to a clear category—which is what image recognition is—isn't always easy. This is especially the case when experts have given fantastical names to a large number of tiny subcategories that could just as easily have been called something else. Things become even more complicated if a child points off in a general direction, excitedly crying out, "Mommy, look, see? What is that?" or if the coveted object lies half-concealed by a beachball. These are the same challenges facing the image-recognition systems so urgently required by industry and for self-driving cars. If a robotic arm is going to pick up a small tool or a self-driving car is supposed to spot a child about to chase after a ball, both devices need one or more cameras, as well as software that tells them what lies where in the surrounding environment.

It also just so happens to be an area in which computer systems have improved drastically in recent years. Here too, it was a contest that brought together a wide range of working groups: the ImageNet Large Scale Visual Recognition Challenge (ILSVRC).[2]

What was actually being measured, though? The data set was made up of 1.2 million (!) images, each of which could be assigned to one out of a thousand possible categories. Every challenge was composed of a training set, or a set of images with known categories for the machines to learn

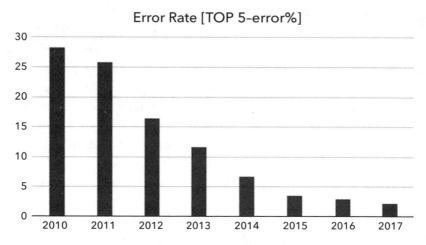

Figure 33
Between 2010 and 2017, the error rate for image recognition by AI fell from 28 percent to just over 2 percent (results from the ILSVRC, Classification and Localization Division; see https://image-net.org/challenges/LSVRC/index.php).

with. As with the Netflix contest, there was also a test set with known categorizations. For each image, the algorithm was now supposed to give up to five categories (out of a thousand) into which the image could be sorted. An image counts as properly identified if one of those categories matched the category to which the image had been assigned.

If your mind works anything like mine does, at this point you might ask whether this sort of approach doesn't make things pretty easy on the computer. If a computer thinks it detects (a) a sunset, (b) a face, (c) a toothbrush, (d) a fish, and (e) a television in an image, and one of them is correct, did it really win the contest? It wasn't all that simple for the machine, however. Many images contain more than one object, and it is often unclear which is supposed to be the main one; it's not always in the middle or taking up the most space.

If you'd like to see the images for yourself, there's a list of the thousand categories that were supposed to be learned at http://image-net.org /challenges/LSVRC/2012/browse-synsets, all of which come from a larger data set called ImageNet. Each category in ImageNet contains a different number of images, some of which are ridiculously difficult. The category Dungeness Crab, for example, sets an image of a living specimen next to a beautifully arranged, neatly carved up crab where all that's missing is a squeeze of lemon. In another image, a fairly pale tourist stands half-naked on a beach holding up two discoveries that are only marginally visible.[3]

If you click through the images, you'll find that some have been horribly miscategorized. The category Foot Ruler, for example—a ruler that is a foot long—also includes a series of images of feet with no ruler at all, but which are evidently a foot long. The folder also contains a tremendous number of images of fish laid out along a ruler, probably to measure them for a contest. The ruler itself is barely visible. In this case, you would be entirely justified in doubting whether this image was assigned the right central category, especially when anyone can see it's a fish—pardon me, an American brown trout with an auditory canal on its left-hand side and a spotted tongue. As everyone knows.

Against this backdrop, all you can do is tip your cap to what machines have managed to accomplish with image recognition. When you consider that they are spotting objects that are themselves hidden and may not even play the leading role in an image, and what's more may contain highly specific categories like different dog breeds, it's a wonder that machines now only classify 2.5 percent of images incorrectly. In 2010 it was still at 28 percent.

But wait a minute—I still owe you a comparison to human labelers, don't I? I'm sure you're imagining that a small army of testers were recruited to solve the same task as the machines. For each image testers would select five potential matching categories, after first having learned the thousand categories possible by heart. But who's going to learn a thousand categories by heart? Good question! The contest's organizers initially looked to outsource the job to Amazon's Mechanical Turk, as often happens with microjobs. The Mechanical Turk is a platform on which registered users can accept microjobs, for which they receive a small amount of compensation in return; the low wages make it more a way to pass the time than a source of gainful employment. In the case of the competition, however, the complexity of the task meant the human testers weren't able to classify the images well enough. The task was thus handed over to two experts who received an initial batch of five hundred images and their proper classifications for training purposes. One of the experts classified 258 images before throwing in the towel; he had misclassified 12 percent. The other held out considerably longer: he classified a total of 1,500 images, at a rate of error of 5.1 percent.

And that's it—the human standard to which all other systems have been compared since. In other words, the so-called and frequently discussed "human rate of image recognition" rests on the accomplishments of a single person with a brief period of training.

It happens quite often in science, by the way, that findings and results take on a life of their own, as it were. It would be much nicer if the single person mentioned in the original article did not in the course of time become *the* human rate of image recognition, as articles cited and copied from yet other articles were in turn copied and cited. Still, it would hardly make a difference in this case if one were to consult someone who had received more or better training. The outcome stays the same: machines are now good enough to recognize the objects an image contains. And if an error rate of 2.25 percent is enough to make you leery of self-driving cars, don't worry: the systems in such cars have the advantage of seeing a great number of images of each object that appears while driving. This lets them make improvements; upon drawing closer, what from a distance looked merely like two strips of light is quickly registered as the strips on two jogging shoes worn by a woman pushing a bicycle across the street at night. But that's another story for another time. It's still the case that for a number of categories and images, a largely sensible and impressive contest advanced image recognition to the point where today it has become a viable option.

## THE ETHICS OF COLLECTING DATA AND
## ESTABLISHING A GROUND TRUTH

Nonetheless, there's still one point to make here about the ethics of data collection. I've already discussed situations where image copyright holders weren't asked whether their images could be used for image recognition. In general, the same held true for the ILSVRC. ImageNet itself only displays smaller versions of the original images, called thumbnails, and each photo URL comes with a warning, "Images may be subject to copyright," which is sufficient in the eyes of the law. It's also enough for whoever is training a machine to simply go down the list of URLs and use the information on the screen. Real, permanent download isn't necessary for learning; showing the colored pixels on one's own screen is sufficient. I would have liked to include a couple of these pictures for you here, but I discovered that the licenses under which they were uploaded don't allow it without permission.

Ready for another fun fact? Google is quite good at image recognition and was often among the leaders in the ImageNet challenge—and it probably did so with your friendly assistance!

Huh? You were never asked if you wanted to participate, you say? Right. Well, at some point or another, you wanted something from a website that wasn't sure whether you were really a human or not. I, on the other hand, can be fairly certain that I'm talking to a human right now because you are sitting with a book in your hand, reading it. To determine this for themselves, digital services like to use what are called *captchas*, from a program developed by Luis von Ahn that stands for "completely automated public Turing test to tell computers and humans apart." Captchas are little puzzles that are difficult for machines to solve but easy for humans. Ultimately, they serve as a security measure by checking whether an entry in a form online comes from a person (you) or a bot that is being misused. *Bots* are small programs that surf the web on their own and behave similarly to people.

Figure 34
A captcha-like puzzle showing the letters *smwm*.

The best-known of these puzzles for telling people and bots apart has to be the distorted, hard-to-read letters that you have to type in for websites with an extra level of security—when confirming a payment, for instance. In the early years, captchas were created especially for this purpose. Figure 34 shows an example.[4] In a TED talk from 2011, Luis von Ahn estimated that internet users worldwide were spending five hundred thousand hours every day solving this type of invented captcha.[5] This led von Ahn to consider whether all this time couldn't be better spent, and he founded the company reCAPTCHA. It used real-world material instead of invented questions: words in photographs that weren't sufficiently recognizable to computers, for example, but fairly straightforward for humans. Another task involves selecting out of a series of images only those that show a bicycle or a bridge, for instance. Especially for your benefit, I pestered people at the image database Pixabay for so long that they finally relented and sent me the reCAPTCHA you can see here. Recognize it? Essentially what you are doing is classifying an image; you are working out a ground truth that can serve as the basis for training a machine. At this point, Google got involved and bought reCAPTCHA. Since then, a small script on the websites of many online services has analyzed your behavior. If the script takes you for a person, you only have to click the little "I am not a robot" box. If you are behaving like a robot, however, you first encounter a modern—and useful—reCAPTCHA puzzle.[6] It's useful in the sense that it allows Google to improve its Google Books service, Google Maps, or Google Street View.[7] These days, it's anyone's guess how many hours total are spent daily solving captchas and reCAPTCHAs; Luis von Ahn's estimate of the number of hours spent solving captchas might already be trumped by reCAPTCHAs. Google itself has talked about hundreds of millions of reCAPTCHAs being solved every day.[8] There's one more fun fact I just can't keep to myself, and that's a short video on YouTube in which a robotic arm demonstrates its own approach to the "I am not a robot" test.[9] Clearly, we need to come up with far more difficult tasks to be able to distinguish between people and machines!

AT A GLANCE: IMAGE RECOGNITION

In summary, it can be said with some certainty that going forward image recognition will be done by machines. That is the case for a couple of reasons in particular:

1. People can only become experts in a small number of categories. One person knows thirty thousand knitting patterns, while another can tell every type of seal apart. But there is no human alive who can distinguish motors, mushrooms, seals, and knitting patterns with the same level of

precision. Machines are capable of handling as many categories as they like, with the caveat that each new category must be learned from the ground up when it is introduced.

2. In the coming years, the image databanks behind all this and their ground truth—that is, images that have already been classified—will only improve and grow more complete. At some point, there will be enough high-resolution, labeled images of every kind of mushroom, motor, knitting pattern, and animal species from every perspective and against the most varied backgrounds to perfect machine precision.

This presents an ideal type of situation for learning systems. Over the years and with the voluntary collaboration of many millions of people, a foundation of data is being laid that will make truly comprehensive training possible. In just a few years, people will be able to conduct automated surveys of biodiversity, for example—an important building block for continually and affordably monitoring the consequences of human interference in important ecosystems. Image recognition, of course, also serves as the basis for all sorts of autonomous machine behavior—as well as video surveillance. For just as in time there will be enough images of types of animals and technical artifacts, many images of nearly every one of us will also be found within image databases.

A second example of an AI system that functions at least as well as human experts was designed for a medical study, to spot a specific form of skin cancer.

PEOPLE VERSUS MACHINES IN CANCER DETECTION

Everyone in my family is pretty light-skinned, though that never seemed to stop my father from running around the tropics without suntan lotion on his research trips. As a result, today his many freckles and skin spots bring him regularly to the dermatologist, who on his last visit there promptly discovered something that had to be removed. Would it have been better for my father to place his trust in a computer instead, however? Could it have detected the skin alteration earlier?

We aren't quite there yet. Still, in August 2018, a highly promising study showed that by using images, a neural network was better able to detect a malignant tumor than fifty-eight dermatologists![10] Similarly to the risk-assessment systems we've already discussed, the machines were trained to identify benign and malignant forms of melanoma (skin cancer) using images, then assign a risk score to each. Essentially, it was possible to sort the images according to their risk of cancer.

To compare between people and machines, human experts in the field were also asked to categorize the images. Participants were only able to choose cancer or not cancer, however—a method that is also quite common and doesn't put people at a disadvantage. It is simply much harder for us humans to sort a hundred images according to risk than it is to decide for each whether it belongs in one or the other category. On average, the experts caught seventeen malignant cases out of twenty total. Out of eighty benign cases, on the other hand, they classified an average of twenty-three incorrectly. When it came time for the machine to sort the images, the study's authors now set the threshold value such that the machine would also have recognized seventeen of the malignant tumors. In other words, they proceeded from the images with the highest calculated risk value to the lowest until they were left with a total of seventeen images that really did show a tumor. The question posed by the team of researchers headed by Dr. Haenssle at the German Cancer Research Center was whether at this point the machine would falsely classify a greater or lesser number of images as positive (malignant). And here the computer performed much better, classifying only fourteen benign tumors wrongly as malignant, just over half as many as their human counterparts.

If the machine had decided, then out of a hundred patients, nine fewer people would have gone home with an unsettling diagnosis. It's a small consolation that skin doctors who had more than five years in the profession did in fact make better decisions on average than the beginners, but all groups tended to prefer one operation too many to one operation too few. Nor can anybody hold it against them: a malignant tumor that has been overlooked carries far graver consequences than the relatively negligible cost, discomfort, and possible side effects of a minor skin operation. Aside from the concrete consequences for the patient, a doctor might also expose herself to disproportionate financial risk: compensation claims in the case of an overlooked tumor are both more likely and involve much greater costs, even when absorbed by the doctor's liability insurance.

THE ETHICS OF THE UNDERLYING DATA

In the study, the authors note that their algorithmic decision-making system was based only on a single, rather small set of data made up primarily of light-skinned patients. This is an issue that's creating a lot of headaches for many image-recognition systems at the moment; one at Google failed completely at recognizing dark-skinned people, for example. In an email to *WIRED* magazine, a spokesperson for the company wrote that "image labeling technology is still early and unfortunately it's nowhere near perfect."[11] In a TEDx Talk that has since been viewed over a million times, Joy Buolamwini, a researcher at the MIT Media Lab, has called attention to the fact that her dark skin is not even detected by common facial-recognition software.[12] In response, she founded the Algorithmic Justice League, an NGO that fights for discrimination-free software. Dark-skinned hands are also often overlooked by image-recognition software—the sensors in soap dispensers, for example.[13]

These problems can largely be traced back to faulty and incomplete sets of training data. Data in the medical field, for example, is known for relying heavily on studies of geographic areas that are more well-to-do. This in turn favors insights about treatment and medication predominantly for light-skinned people. Viewed historically, the data is dominated by light-skinned men.[14] Such databases urgently need to be rounded out before they are used for training—that, or their use must be explicitly restricted to cases that have been demonstrably tested. Overall medicine is actually one field where enough data of sufficient quality could be gathered over the years—if society wants that. The training data necessary to create neural networks, including a carefully elaborated ground truth, could be obtained by pathologists working in sufficient numbers and quality.

In the present case of cancer detection, a system would have to be built that was actually able to recognize melanoma for all skin types—that, or one system for each type of skin. If it worked, the division of labor could benefit dermatologists, especially in cases where no treatment was necessary, which might in turn help balance out the asymmetrical costs between patients and doctors in the case of wrong judgements. Kun-Hsing Yu, another doctor who has focused on the question of algorithmic discrimination, isn't worried about machines replacing doctors: "It's more like doctors who don't use AI will be replaced by those who do."[15]

But why is it exactly that cameras paired with image-recognition software are now better at recognizing skin cancer than doctors? Isn't that something we could test using the scientific method? In the next section, I sketch how we might test the hypothesis that machines hold an advantage over people in similar fashion to the hypothesis that apoptosis is beneficial for yeast.

## WHERE MACHINES ARE ABLE TO LEARN

The two preceding examples showed that there are algorithmic decision-making systems out there that can either replace humans or provide them with meaningful support. Yet it's also become clear that AI with a learning component is only ever plan B. That's because the decision rules an algorithm learns will always be dependent on the concrete training data it is fed, as well as the many parameters selected for the method. What's more, most of the methods of machine learning remain incomprehensible to humans. That applies especially in the case of the statistical models learned by neural networks. For example, while it would be possible to trace and understand the end result for each individual input, it's practically impossible to predict the general behavior of even slightly different inputs. This is where the issues raised with Knuth and Pratt's classical typesetting system come into play: complex software behaves in a way that simply can no longer be described in the abstract. If it is in fact possible to list the decision rules in a situation clearly and in a structured way, the preferred method is usually an expert system. An *expert system* assembles human-made rules into a structure—a decision tree, for instance (one that isn't learned, however, but designed explicitly by people), or a data bank. The actual decision algorithm then goes through the rules with whatever new data is provided. This means that at any point, humans can follow all the decision rules being applied, as well all the decisions that end up being made. This is a first type of situation in which software systems with a learning component wouldn't be used.

The second type of situation where software with learning components wouldn't be used is one that can be modeled by a mathematical problem for which a classical algorithm already exists. Depending on how long it takes the algorithm to calculate the solution, the algorithm is given preference as it does actually find the *best* possible answer. Remember that most algorithms used in machine learning are only heuristics, sets of instructions that attempt to find a solution but can't guarantee that they will find the best one.

How do we become convinced that the answer a heuristic has found actually has legs? The methods of machine learning search for correlations. That means they search for properties that very frequently appear alongside the property that is to be predicted (and conversely appear less frequently if a data point doesn't include the property to be predicted). Initially, however, they can't tell us anything about whether or not such correlative properties are in fact the cause of the property in question or influence it in some other way. In terms of the scientific method, this means they only reflect the first step—*observation*. To follow the method fully, one would first have to develop a hypothesis based on multiple observations and test it experimentally, until multiple experiments yielded a theory. Only after this theory was then substantiated by repeated rounds of prediction and observation would we begin to talk about a fact. But with machine learning, the hypotheses established by the machines are manifestly never inspected for causality. So why are machines nevertheless still allowed to make decisions?

In 2008, Chris Anderson, then editor-in-chief of *WIRED* magazine, spoke in this context about an "end of theory" that would make the scientific method obsolete. In light of the mind-blowing quantities of data, a new approach was needed, and that was to be found in mathematics. His exact words were these: "There is now a better way. Petabytes allow us to say: 'Correlation is enough.' We can stop looking for models. We can analyze the data without hypotheses about what it might show. We can throw the numbers into the biggest computing clusters the world has ever seen and let statistical algorithms find patterns where science cannot."[16]

This captures quite well the sort of gold rush mentality that prevails among the ranks of technological enthusiasts. And in part, they're right; in many cases, machine learning is able to achieve results that either exceed human capacity or are at least good enough to make them more efficient. In many cases, they will improve our lives. Yet that will only come to pass if we're able to inspect the quality of those results. Sometimes we can do that through *induction*, or by drawing general conclusions based on special cases. Every day, search engines make millions upon millions of predictions that are either reinforced or corrected by users. If the decision rules guiding these predictions manage time and again to place meaningful suggestions among the first ten search results, this bolsters our trust in the underlying statistical model. The more often an observation like this gets made, the safer it is to conclude that it will continue to be like this in the future based on the fact that up to now it has always been the case. Still, there's always the risk that something totally different will occur the next time—something for which Nassim Taleb makes a highly compelling case in his book *The Black Swan: The Impact of the Highly Improbable*.[17] This means theories are much more reliable when the causal relationship between the input data and the facts that are to be predicted is known. And if there is no relationship to begin with, then the attempt to learn something on the basis of the training data is doomed from the start.

Essentially, machine learning can succeed when the following conditions are met:

1. The *amount of training data* (inputs) is both large enough and of sufficient quality.
2. A truly high-quality *ground truth* (i.e., what should be predicted, or the output) exists *that can easily be made measurable*.
3. *Causal relationships* exist between the input and the output that is supposed to be predicted.

The algorithms of machine learning are plainly superior to humans in these cases because they

1. can search for correlations among quantities of data of nearly any size,
2. can search for a great number of different types of correlations, and
3. can profitably incorporate even weak correlations into the statistical model.

Particularly in cases where only weak correlations exist or it's initially unclear which input data even influence the output that is to be predicted, access to vast quantities of data (big data) can help iron out the weaknesses. That in turn means that many of the questions we'd like machine learning to answer basically depend on large quantities of data.

The results of machine learning gain credibility if the additional conditions are also met:

1. Enough is known about the causal relationship between input and output that a clearly definable quantity of input data exists about which all involved parties can easily agree.

2. There is as much feedback as possible for both types of errors (false positive and false negative decisions). This allows quality to be measured continuously and for the statistical model to improve dynamically.
3. A clearly definable quality metric exists on which all involved parties can easily agree.

Especially with human behavior, predictions first became conceivable with the combination of big data and the algorithms of machine learning. Where before individual parameters and their impact on human behavior had to be laboriously worked out via lab experiment, big data can now help set up statistical models for many situations. This is especially the case in the commercial world, where it isn't necessarily the goal to make the best models possible or ensure they only include causal relationships between their input data and the prediction at the end. A model still serves a purpose, after all, even if it manages to predict with just a bit more accuracy which piece of news might be of interest, which ad someone might click on, or which product they might buy. In doing so, it's easy to look past the fact that commercial uses of machine learning succeed with human behavior because they meet the conditions set forth previously—namely:

• Incredibly large quantities of data on human behavior exist on the internet, even if that data is not available generally but only to a select audience.
• The ground truth is easy to measure: "clicked on the news headline/ad" or "bought product" is a simple, binary observation.
• The predictions can thus be tested for their caliber continuously, and the quality of the statistical model can be improved within the realm of what's possible.
• A clear quality metric exists, which is the financial gain in whatever currency you're working in. The absolute quality of the statistical model is irrelevant for increasing profit; it has only to be a relative improvement over the models used to date.

Yet broader societal damage can be inflicted even under conditions that are ideal for machine learning. This has been written about and discussed at length, whether in terms of discrimination brought about either directly or indirectly by algorithms, radicalization through digital social networks, or harm caused by surveillance algorithms, to cite just a few issues.[18] What concerns me the most, however, is the overly optimistic notion that algorithmic decision-making systems' success in the world of digital commerce can be applied just like that to totally different realms of human behavior. As the remainder of the book shows, the baseline conditions for lending credibility to machine prediction are lacking in many areas where machine learning is either being planned or already in use. That means the systems require meticulous technical supervision, and should be prohibited in cases where the conditions either aren't right for machine learning to succeed or its use poses too great a potential for harming society as a whole.

After quickly reviewing the ABCs of computer science, we go on to take a second, deeper look through the algoscope to explore the weak links in the long chain of responsibilities. We consider where you can participate in creating algorithmic decision-making systems and which systems require which type of supervision. With that, you will finish the book poised to obtain all the information you'll need to classify and help shape the algorithmic decision-making systems used by your school or university, place of work, or government.

## ARE WE LITERATE YET?

We've now completed our extended journey through the ABCs of computer science—I hope it wasn't too dusty for you backstage! In the hundreds of lectures and dozens of interviews I've given over recent years, one question I'm asked time and again is what we can do about cases where algorithmic decision-making systems do end up leading up to discrimination or other errors. To answer this question, we have to go backstage once more and see for ourselves exactly what pulleys and levers let algorithms solve problems—or cause them.

I hope by now you feel you've achieved literacy when it comes to the basics of machine learning, which at present is the most important piece of AI. Just to play it safe, let's go over it one more time. First, I emphasized that in classical problems, *modeling* the everyday situation—that is, simplifying it so that one of the many *classical algorithms* in existence can calculate an answer entirely without machine learning—represents the single most important act of design. The examples I gave for the sorting problem and shortest path problem showed just how many everyday questions can be reduced to these two, something which applies for all algorithms and is their secret superpower. Algorithms' usefulness lies in the fact that it doesn't matter to them what the data itself means; they work exclusively on the basis of the data's numerical value. Nevertheless, even in the case of classical algorithms, the *OMA principle* still applies. Interpreting algorithmic results can only ever be meaningful in the context of their *modeling* and the *operationalizations* necessary for it.

*Machine learning* attempts to identify connections between input data and an observed result (output). The correlations algorithms discover in a *set of training data* are stored in the form of decision rules in one of a number of possible structures (*decision trees* or *mathematical formulas*, for instance, or *support vector machines* or *neural networks*). Doing so involves setting and adjusting any number of *hyperparameters* (knobs, pulleys, and levers). One can also change the quality of the prediction by changing the input data (*feature engineering*).

The quality of a given prediction is measured by a *quality metric* using a set of *test data*. The quality metric is what fundamentally determines the direction in which the hyperparameters should be adjusted. This makes it critical for the quality metric to match to whatever *social situation* the algorithmic decision-making system will be used in. Here is where decisions that are largely ethical in nature get made.

Once the structure has been built, we are dealing with a *trained statistical model*, in computer science speak. New data is then led through the statistical model using a second, very simple algorithm, which results in a *decision that takes the form of a single number*, which can represent either a *classification* or a *risk assessment*.

*Big data* plays an important role in all of this because most of the algorithms in machine learning are *data-hungry*. This doesn't depend simply on the method, however, but on the situation that is to be learned. If only *weak correlations* exist or it is unclear *which input data correlate with the output*, resilient statistical models can only be trained using large quantities of data, and only to a certain extent. Constantly and dynamically weighing predictions against observed output (e.g., human behavior) can shore up our confidence in the statistical model's usefulness.

## THE PATH TO BETTER DECISIONS,
## WITH AND WITHOUT MACHINES

Machine learning, then, is only plan B. But unfortunately, there's often no plan A either, because many of the problems waiting to be dealt with can't be solved by classical algorithms. There is, of course, an alternative: for humans to continue to solve these problems themselves, or search for answers alongside machines. This makes it all the more important for societies to start thinking and deciding now about what we mean when we say "good answer," because that's a decision no one can take from us.

ALGORITHMS, DISCRIMINATION, AND IDEOLOGY

I'm in the Paul Löbe Building in Berlin's government district, at one of the monthly meetings for the Enquete Commission on Artificial Intelligence. The conference room is relatively modern; everyone is seated at tables arranged in circular fashion, giving all thirty-eight members of the committee clear lines of sight to each other. I peer down and am sad to discover that the coffee I purchased at the kiosk outside, as underwhelming as it was, has long since vanished. We're in the middle of a question and answer session following a report on machine learning and discrimination, one part of our work in drafting a series of recommended actions concerning AI by mid-2020 for the Bundestag. Suddenly and without warning, one of the committee members gets huffy: "Firstly, I must say that the hopes expressed at the outset of these meetings, about the possibility of our holding a dialogue on the topic without ideology entering the picture, have proven quite naïve. As has become readily apparent, the same ideological issues manifesting themselves in other fields of politics are also on display here, perhaps in even more virulent form."

Huh? Now where on earth did that come from? I was thunderstruck; up to that point, our work up had been quite fact-based. But let me give you a bit of background. The Enquete Commission was set up by the Bundestag in October 2018 to develop a set of recommendations concerning social responsibility and the economic, social, and ecological potential of AI. The group included nineteen members of Parliament, with party representation proportional to the composition of the Bundestag at the time. That meant seven members from the CDU/CSU, four from the SPD, and two each from the AfD, FDP, Left Party, and Alliance 90/the Greens. Each party faction additionally sought out a number of experts in the field, resulting in a motley crew of philosophers, sociologists, and political and computer scientists, as well as representatives from businesses, unions, and various institutions (among them Initiative D21, the Stiftung Neue Verantwortung, and D64). When Anke Domscheit-Berg, a former member of the Pirate Party now caucusing with the Left Party as an independent, asked whether I would come on board as an expert, I leapt at the chance.

The commission was formally inaugurated on September 27, 2018; Bundestag President Dr. Wolfgang Schäuble gave a short speech at the event. He spoke about how for many, "artificial intelligence . . . [ranks as] the new magic spell of technological progress," listing the many things intelligent software would be capable of in the future. Among the positive aspects was machines' coming ability to "compose verses," though he warned they could also be used "to reward and to punish."

These two aspects seem to me to be what most concern us as humans: that AI will create poem and poena alike (where *poena* is the old English for *punishment*). "Creating poems" serves here as a stand-in for our most elementary activities as humans and the worry that one day machines will deprive us of them. This is a concern we will have to respond to at a societal level through labor, educational, and social policy. When it comes to AI "creating poena," technical oversight is primarily concerned with ethical and social aspects. What, for example, might a German or European approach look like in this case?

Aside from technologies that replace human decisions, the US, as a point of comparison, also relies on platform business models. In recent years, such models have exhibited serious side effects, manipulation by third parties, and, most recently, data protection scandals.[1] China also has access to massive quantities of data and has been able to develop a series of state technologies based on them. In particular, different versions of a "social credit system" intended to encourage "positive behavior" among Chinese citizens are currently being tested. All of them evaluate citizens' behavior and rate it with a single number: if you pay off your credit card, that's a plus; if you don't keep your children well-fed and clothed, it's a minus. Germany and Europe may want to use the technology in a secure manner that is light on data and keeps sight of the individual person, but what exactly is that supposed to look like? The Enquete Commission was formed to consider just this kind of question. In what follows, I'd like to introduce you to a couple of possible answers, some my own and some the commission's.

Our first sessions in the fall of 2018 were largely spent defining terms. What do we mean by artificial intelligence, and what is machine learning?

What can and what can't it do? In December, we then invited a number of government ministries to explain more about the goals behind the German national strategy for AI, which had been published the month before.[2] Since then, the commission had focused on developing a series of concrete recommendations complemented by a set of ethical questions.

Most of our meetings followed the same pattern: experts in their respective fields who either sat on the commission or had been invited to speak began by providing input (this part is public, by the way—you can find videos on the web), followed by a question and answer session.[3] In January 2019, one member of the commission, a professor of computer science named Hannah Bast, gave a report on machine learning's ability to draw *analogies* from text. For instance, when algorithms are shown texts that feature the word pairs Germany-Berlin and France-Paris often enough, they are able to draw analogies such as Berlin is to Germany as Paris is to France. This is of potential interest for any number of issues, even without explicitly mentioning the concept of a capital city. In other texts, the same type of algorithm was able to learn different job titles for men and women, through analogies like this: man is to waiter as woman is to waitress.

Yeah, right. The same algorithms we've seen functioning so flawlessly up to this point could just as easily be learning instead that man is to doctor as woman is to . . . nurse! At least in Germany, where the share of female students entering medical school passed the 50 percent mark more than ten years ago, such a stereotype is antiquated.[4] In this case, the inference is *factually incorrect*.

As I'll discuss in the next section, this isn't the only case where changes in today's work world conflict with reflections of older stereotypes in AI systems. In her report, Professor Bast went on to mention that other methods existed for straightening out this sort of data before further processing. And while one half of the conference room wanted to learn more about them, the other grew indignant about technology being dressed up in ideology, as the quote at the beginning of the chapter shows.

As it is, questions about whether and how to deal with stereotypes, bias, or discrimination in data and the algorithmic decision-making systems that result take us out from backstage. They really aren't something for a data scientist to decide but societal issues that demand a response at the national, even international level. Before you can help make those decisions, my dear reader, we should take one last look back at the whole operation—from the balcony down onto the entire backstage, so to speak.

## HOW DISCRIMINATION FINDS ITS WAY INTO ALGORITHMS

The United States has the highest rate of imprisonment out of any country in the world. In concrete numbers, out of every one hundred thousand US citizens, more than 650 sit behind bars.[5] What's more, the rates of imprisonment are extremely uneven. Residents of Hispanic origin are three times as likely to be imprisoned as their white counterparts; African Americans are six times as likely.[6] The links between the true rate of criminality and the rate of imprisonment are incredibly complex. It's not just about how often people are subjected to "stop and frisk" policies on the street; it's also about which kinds of criminality tend to be prosecuted (drug crimes, for example, as opposed to white-collar crimes like tax evasion); who can afford bail and who can't; and the different types of sentencing people receive. The ACLU points out that many of these individual steps perpetuate unjustified forms of discrimination that lead to the rate of imprisonment being a distorted reflection of the actual rate of criminality. In 2011, this prompted the organization to call for the use of algorithmic decision-making systems in every phase of criminal proceedings, as a way of helping to reach decisions that were more objective and free of discrimination.[7] But is it really all that simple? Do computers never make mistakes? Are they totally objective? In what follows, I point out the many places

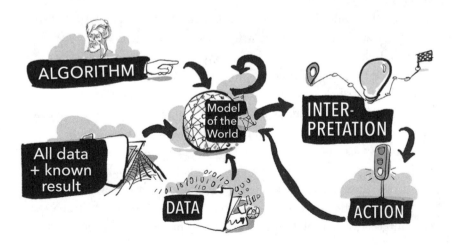

along the long chain of responsibility that can result in decisions that wind up causing discrimination.

In English as in German, the word *discrimination* is nearly always used in a normative and moralistic sense, as an unjust preference or prejudice shown to different groups of people defined by characteristics such as sex, age, faith, or ethnicity, among others. Now, algorithms and the heuristics of machine learning are built up around finding differences among groups with regard to whatever behavior they are being trained to recognize (the output). It's in their coding DNA to separate groups from one another, or *discriminate between them* in the neutral, technical sense of the term. Nor is there any problem with that, so long as it doesn't involve certain specific forms of discrimination, or characteristics protected under a country's basic constitutional rights. In Germany, those characteristics include ethnic background, gender, faith or worldview, disability, age, and sexual identity. Particularly serious instances of discrimination based on these properties might include discriminating when hiring employees or choosing members for a works committee or another employee association. Unjustified discrimination is similarly prohibited when it comes to wage agreements, any state service, education, and public infrastructure. All this is to say that if an algorithm is either responsible for (or an essential part of) the decision-making process in one of these areas, its decisions must be inspected for forms of unjustified discrimination.

Sometimes it can be difficult to say whether *bias* in a decision represents a form of discrimination. In Germany, there absolutely exist jobs for which men can be given preference. Those who want to enter the police force, for example, must on average demonstrate a certain body weight that, statistically seen, is met more often by men than by women. For a while, the state of North Rhine-Westphalia had different minimum weights for men and women to account for that very fact. During that same period, a smaller male applicant who was still above the minimum weight requirement for women brought a suit against the law.[8] The constitutional court in Gelsenkirchen ruled that it was permissible to fix different weights for men and women on principle, but they had to be particularly well justified—something that wasn't the case in that suit. One interesting aspect of the judgment, however, is that fundamentally the court also viewed it as objectively justified to exclude higher percentages of women through minimum respective weights in order not to lose too many male applicants.

The court didn't set out any guidelines for what those weights might be, of course, but left the decision to the state of North Rhine-Westphalia. This meant there was no conclusive opinion established on two conflicting societal goals, let alone their *operationalization*: making equal rights a reality on the one hand, and the state's justified interest in fielding a sufficient number of suitable applicants for the police force on the other. Now let's imagine that simply by combing through the properties of successfully employed policemen and women, an algorithm discovered that success did in fact correlate with body weight. Would that be the socially desirable answer? Whatever the case may be, algorithms and heuristics play an ambivalent role in this context. For one thing, they are capable of uncovering existing distortions and thus bringing them to light in the first place. Were appropriate measures then taken to balance out those unjustified forms of discrimination, data science would even have contributed to creating a more just society. Then again, naively and directly applying learned decision rules that contain bias can preserve, even strengthen socially undesirable forms of discrimination.

Discrimination thus represents one of the thorniest problems out there when it comes to using algorithmic decision-making systems. What's more, discrimination can seep into ADM systems at nearly any point along the long chain of responsibility or be reinforced by their use. Yet as before, only very few steps in the chain require any form of technical knowledge to identify actual discriminatory decisions. *Every* step along the way, by contrast, requires some form of social consensus.

DISCRIMINATION IN THE DATA

Let's take as an example the widespread aspiration for businesses to be able to tell from looking at written applications which candidates will prove successful hires. Some companies also enlist chatbots or video platforms in their application process so that systems running behind the scenes can evaluate how good of a match the applicant will be. Throughout all this, obviously, runs the hope that the software will operate free of discrimination, or at least with less discrimination than if the decision were left to humans. In 2014, Amazon began to develop an automatized system for reviewing applications, using the resumes submitted from the previous ten years as input.[9] As it turned out, in that period most of the successful candidates were men. It's not public knowledge whether men were hired disproportionately to the respective percentages of male and female applicants. But it is public knowledge that today, a mere one out of every five employees specifically at technology companies like Apple, Facebook, Google, and Microsoft is a woman.[10]

Although the heuristic used to train Amazon's statistical model didn't receive the gender of the applicants as input, it still managed to discover properties that correlated with gender. If those properties surfaced in the applications, the person received a lower rating. Mentioning membership in a women's chess league on one's resume was rated negatively, for example, as were referrals from women's colleges, presumably by the same logic. It's the sort of thing that can happen when an algorithm determines that this sort of person was a less successful hire in the past.

The developer team at Amazon patched up both instances when it became clear what was going on, though no one could tell for certain if that meant every last issue with discrimination had been cleared up.[11] This is something that's actually quite difficult to be sure of as a selected input can create discrimination in very subtle ways. As one person on the inside reported, the review software also took applicants' self-descriptions into account when judging whether a person was fit for a job at Amazon, either from their cover letters or websites. Self-descriptions are another context where systematic differences based on a person's origin and gender often emerge. The decision to include them, then, could also create a situation where whatever had worked frequently in the past was reinforced disproportionately through the use of machine learning, resulting in a continued monoculture. In the end, the project was discontinued.

This example shows unmistakably that when the selection process has already led to a biased outcome regarding a sensitive property (in this case, the applicant's gender), an algorithm is able to discover that by correlating it to others, even without knowing what that property is. In this case, a fairly unambiguous correlation existed between various leisure activities, a handful of colleges, and gender. If the resulting statistical model is then reapplied uncritically, the bias may well grow stronger.

The difficulty in the case of such biases lies in their interpretation. Were the groups that received different ratings equally well suited for the jobs? Was an unjustified form of discrimination really at play? With Amazon, all we can say is that the machines uncovered a form of unequal treatment in the data, which they then passed along.

In brief, if discrimination (whether justified or unjustified) was present beforehand, machines will learn it in their training.

## DISCRIMINATION BY LACK OF DATA

Discrimination may also appear in results if data from specific groups is lacking. We've seen this before with image recognition. In that case, systems would require representative data from all kinds of people to be able to recognize hands and faces with different skin colors, for example, or differentiate between melanoma and harmless liver spots. Speech systems likewise need input from a large number of people, especially so those who speak with accents, dialects, or speech impediments can be understood just as well as those who don't.[12] In July 2016, sociolinguist Rachael Tatman reported on her blog that the best speech-recognition system at the time showed a statistically significant advantage in recognizing men's speech over women's.[13] When she repeated the experiment in 2017, she was no longer able to detect a difference between the sexes.[14] With that said, she was able to ascertain in a new set of experiments that people who spoke with a strong accent—those from South American countries, for example—weren't understood as well.

Voice recognition may still seem like futuristic stuff, with only limited use in our everyday lives. Yet computer scientists are unanimous in the view that in the not too distant future, speech interfaces will by and large be more common than the keyboard and mouse. There's a wonderful sketch on YouTube with a pair of Scotsmen who are driven insane by an elevator using US voice-recognition software.[15] And it's not just on YouTube: in Australia, speech recognition is used for certain work visa applications to determine whether the applicant's English is good enough. A native of Ireland wasn't about to convince the machine she could speak English![16] The reason immediately suggests itself that the machine wasn't trained with people from Ireland, although they're obviously capable of speaking the language.

Many argue that the lack of data is in itself biased—in other words, that it follows a pattern. Caroline Criado Perez has written an entire book about data that is missing from and about women.[17] In a similar vein, a white paper from the World Economic Forum titled *How to Prevent Discriminatory Outcomes in Machine Learning* brings together several studies that show how people in developing countries, especially women, are notoriously underrepresented due to their limited access to the internet and digital devices.[18]

In this case, it clearly stands to reason that systems can't be trained to incorporate properties for groups of people that do not appear in the data in the first place.

### DISCRIMINATION BY OMITTING SENSITIVE INFORMATION

Surprisingly, discriminatory decisions can also be found when machines aren't given information about a sensitive property but differences in behavior do exist between two groups in a population. To show you how that works, I'll pick back up with the made-up example from our data set on recidivism. You're familiar already with how the data set looks (see figure 35).

It is impossible to draw a line for this data set that cleanly separates the guilty from the innocent. Yet depending on the circumstances, such a line could be drawn independently for women and men. In figure 36, I've

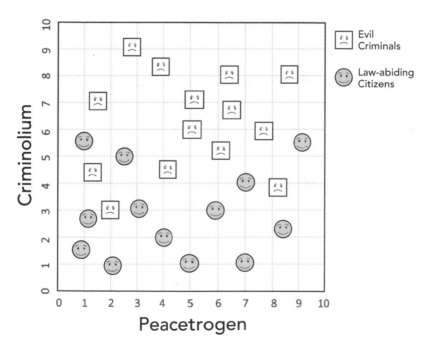

Figure 35
The original (imaginary) data set.

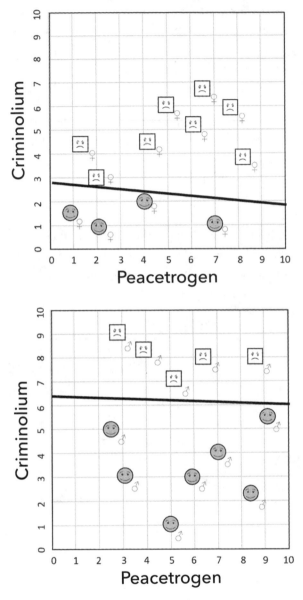

Figure 36
The data set shown in figure 35 has been divided into female (upper chart) and male (lower chart) criminals. An optimal dividing line exists for both sets.

divided up the data set into men and women. Lo and behold, an optimal can be found for both sexes!

If we had divided the data up then, and trained a support vector machine for each, we could have found an optimal decision for everyone. When the data is taken together, it's no longer possible to find an ideal dividing line. As figure 37 shows, a dividing line with the least number of errors puts men at a disadvantage.

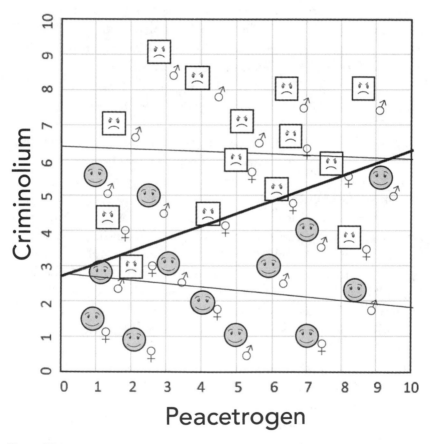

Figure 37
The top and bottom lines show the optimal dividing lines for each respective sex. The black diagonal line shows the optimal dividing line when the data for both sexes is trained at the same time.

Now, the black diagonal dividing line places two innocent men on the side of the criminals, and two criminal women on the side of innocent citizens. Discrimination can thus arise when sensitive properties are withheld from the algorithm if those properties are in fact causing differences in the predicted behavior to begin with!

Even when such sensitive data is included, not all methods of machine learning can make use of it. A support vector machine cannot necessarily incorporate it, for instance, while a decision tree can. As such, the method itself can also lead to discrimination.

Finally, discrimination can also occur when the statistical model is continuously trained; that is, when feedback for predictions is applied constantly to improve the model. These statistical models are also called *dynamic models*, while training is sometimes referred to as *online training*.

## DISCRIMINATION BY CONTINUOUS LEARNING

In 2016, a gentle creature was released into a harsh environment. A chatbot, Tay, had been designed to interact with people on Twitter by learning what they were talking about then volunteering its own opinions. Tay's Twitter account was furnished with the image of a young woman, albeit somewhat pixelated in a nod to the bot's digital origins. A *bot*, as I explained earlier, is software that can navigate through a digital environment on its own. On Twitter, that meant it could like and retweet posts, but also create its own posts.[19] The account announced its arrival with a chipper "Hellooooooo World!!!," but before too long it was accusing President George W. Bush of causing the 9/11 terror attacks, spewing racist and sexist tweets, writing about all the people who deserved to be hated, and how right Hitler had been.[20]

How could something like that happen? Well, the bot was designed to learn from what other users wrote to it. So when an unscrupulous group decided to bombard the bot with unsavory opinions, that was what happened. We have a saying in computer science: "Garbage in, garbage out."

Could Tay's engineers have prevented such an outcome? Now *that* is a fascinating question. Certainly, they could have prevented a handful of the most obvious offenses by blacklisting certain hateful turns of phrase; you can safely intercept any number of insults or death threats made to individuals or groups that way. On the other hand, automatically filtering the text quickly reaches its limit if people intentionally seek to get around it. Human creativity in such pursuits is one of the distinguishing marks of our species, after all.

Automatic filters being complicated is no excuse to not build them into a system, but it does mean that there is no such thing as a perfect filter. And thus—you'll recognize the pattern—it is first and foremost a question for society of how we want to weigh the errors that machines will inevitably make against one another. Is it better for too much or too little to be deleted? Do different topics get different answers?

By this question at the latest, the issue primarily becomes a societal one. Hans Block and Moritz Rieseweick's documentary *The Cleaners* reveals how companies like Facebook and Twitter are still hiring people in countries that pay low wages in order to make these decisions for us.[21] Around one hundred thousand such "cleaners" spend hour after hour wading through violent footage, degrading sex scenes, and child pornography so that those in the "first world" don't have to. What they are exposed to has almost certainly been automatically prefiltered and is likely borderline material that is too difficult for computers to decide on. It's obvious at any rate that such large companies would long since have automated the process entirely had they been able. In this instance, the values of freedom of opinion on the one hand and human rights on the other often lie trapped in a contradiction that can't be solved by simple rules. This speaks volumes about the impossibility of perfecting upload filters, which is a major cause for debate in Europe at the moment.

AT A GLANCE: HOW DISCRIMINATION ENTERS THE COMPUTER

Along the long chain of responsibility there are a number of points where discrimination can seep into the decisions of a trained algorithmic decision-making system:[22]

- If the *data* contains discrimination, either explicitly or implicitly, which the algorithm then correlates with other variables.
- If differences in behavior exist within the population but the *overall data* for several or many of these groups is missing.
- If differences in behavior exist within the population but sensitive information is withheld from the algorithm or heuristic—*one piece of the data* is partially missing for everyone, in other words.

- If differences in behavior exist within the population but the *method* of machine learning isn't able to distinguish between different contexts.
- If systems are released "into the wild" and falsely trained. In this case, the problem is *controlling the input data*.

As you can see, the bulk of problems involving discrimination revolve around the data; in some cases, the method plays a part. Data can be carefully inspected before it is used: Does it already contain forms of discrimination? Are all relevant segments of the population represented in sufficient number? Discrimination can also creep in during subsequent phases, which makes it important to inspect the results as well. That happens using something called *black box methods*. Most of the time, however, the statistical models are too complex to judge their behavior straight away and are thus often called *black box models*. They need to be analyzed instead as artifacts in studies that use black box models, and that often resemble the sort of investigation I pursued with my yeast cells. To look for discrimination, we observe the machine's behavior with respect to gender, race, or any other sensitive attribute.

BLACK BOX STUDIES FOR IDENTIFYING DISCRIMINATION

In 2015, Amit Datta, Michael C. Tschantz, and Anupam Datta conducted an automated study of job postings in which they set up a series of fake Google accounts. Google accounts let you announce your gender at quite a general level; the company argues that it helps to personalize what users see and, of course, allows for specifically tailored advertising.[23] The three computer scientists made some of the accounts male and others female, then had both groups surf the web. The simulated internet users each visited the same websites, proceeded to a job search website, and ended up at an online magazine. The researchers then collected and analyzed all the ads Google showed the users on the last website. They were testing for whether men surfing the web were shown ads at a statistically significantly higher rate than their female counterparts. They were essentially asking the same question I had with my yeast cells as there were obviously no advertisements shown exclusively to men or women; there was no clear result, just a shift in tendency. In their analysis, they reported men as seeing advertisements "related to higher-paying jobs" more often, at a rate that was statistically significant.[24] Although the study in fact only focused on a single ad—a coaching session in order to land what would (presumably) be a lucrative job—it still shows how men and women are treated differently on the internet.

This study is what a typical black box study looks like. In this case, two groups were set apart by only one variable: the gender they gave for their Google account. Just like with my yeast cells, here a scientific trial in which two groups are distinguished by only one property helps to determine whether the software is treating them differently. A human artifact, in other words, becomes an object of the natural sciences.

As it turns out, when it comes to any somewhat complex software, this is the only way we have of understanding the algorithm's general behavior.

Black box testing is a straightforward affair when there aren't too many groups being treated differently by the algorithm. All that Datta, Tschantz, and Datta were looking to find out was how men and women were treated differently, which meant all they had to do was conduct a single experiment with the two groups. In this case, there was only one property with two forms of expression under consideration.[25] Now, however, imagine that an algorithm must distinguish among three different properties with two possibilities each (e.g., more or less than thirty years old, owns or does not own a car, annual income of $35,000 and higher or less than $35,000). Suddenly, you have to simulate eight groups to capture every possible combination of the three properties.

This example shows further that personalizing a service makes it much more difficult to make out the general behavior of an algorithm. Personalization means that an algorithm provides a slightly different service to each user; this will be familiar to you from social media sites, which put together what users see when they log in on an individual basis. Hardly anything is the same for two different accounts, making it difficult for individuals and society alike to check for improper forms of discrimination. This in turn opens the floodgates to personalized pricing; job listings made accessible only to certain individuals, whether intentionally or unintentionally; and certain groups of people being shut out from any number of services. If this were to affect smaller groups within a population, society at large would only take note once it was too late. In this case, then, black box testing has run up against a wall.

Nor is it enough to look only at the decisions made by the machine when conducting a black box test. In the previous example, the algorithm responsible for distributing advertisements made automatic decisions about who would see what. But if a person then comes along and either interprets or acts on the decision once it's been made, then black box testing has to be repeated. What society does with the results of a decision-making system, after all, can either promote greater equality or perpetuate past discrimination.

This is what led the ACLU to change its opinion on pretrial risk-assessment systems. As you'll recall, in 2011 the group was still advocating the use of such systems for every step of the justice system, especially before a case went to trial. In 2018, however, the ACLU joined a number of other civil rights organizations in calling for the systems' abolition, arguing that they hadn't led to the desired decrease in discrimination against African Americans.[26] A black box test of the overall process, composed of the software and all the social actors involved in the criminal proceedings, had led to the insight that the selected means hadn't achieved the end: it hadn't lessened discrimination.

Yet the biggest problem is that what some perceive as a just balancing act, others will perceive as unfair discrimination. That was what happened at the Enquete Commission meeting I described at the beginning of this chapter, when the Bundestag representative abruptly called for the committee to free itself from "ideology."

> On this key point, freedom from ideology won't be possible in either direction. That's because if we want to quantify the discrimination in a set of decisions with respect to subgroups, the question is ultimately how discrimination should be *operationalized*—that is, how to rate the extent of discrimination with numbers. A second step must then determine whether the discrimination that has been discovered is legally or morally justified or not. And what one person will see as "free from ideology" another will see in the opposite light, as ideologically founded—and vice versa. What's more, the problem exists independently of whether it is a human or a machine making the decision.

As we've already seen with the minimum weight requirements for police, it is a question of choosing between (corrective) justice and fairness. You might also think of it in terms of *equity versus equality*. Fairness tries to distribute a limited resource equally: "everyone gets the same amount." Most resources serve a single purpose, however, and if every person isn't able to benefit the same from whatever resource is being distributed, the resulting distribution will be unjust. Take the example of how a teacher might divide her attention between students. Some will need more attention than others, which means that if every student receives an equal amount, then their ultimate success in learning (i.e., level of education) will be unevenly distributed. If, on the other hand, the children's education is viewed as what truly matters, the aim of corrective justice will be for as many people as possible to receive this second resource, while taking into account that the first resource (attention) will be unevenly divided.

This isn't just some academic discussion about competing concepts of justice. It is a highly tangible problem that winds up affecting all kinds of decisions, be they human- or machine-made. When the journalist think tank ProPublica first reported on COMPAS in 2016, it labeled the risk assessment software racist.[27] Why? Because African Americans made up a larger share of the false positive category than the criminal category. That meant the group suffered more from unfounded suspicion than other groups—without a doubt, one aspect of injustice. The company Northpointe disputed the accusation that it had manufactured racist software, pointing out among other things that ProPublica had used the wrong fairness metric.[28] In the computer science community, a different measure applies: namely, every group that lands in the same category must show the same rate of recidivism. That means that if a person is assigned to Category 8, the recidivism rate for his group must be the same as all the other groups in that category. This also makes a lot of sense as a requirement; if that weren't the case, then judges would have to take note of what a given risk class meant for each different subgroup. Unfortunately, this is an either-or situation. Computer scientist Jon Kleinberg, along with Sendhil Mullainathan and Manish Raghavan, demonstrated that it was *impossible to satisfy both fairness requirements at the same time* if one of the groups showed a different rate of recidivism.[29] This means that society must decide for itself which measure of fairness should be optimized. Should a category mean the same thing for all groups? Or should the share of false-positives among groups in the high-risk category be made to match their respective shares within the population, so that the burden (of those wrongly judged) is evenly distributed? In this case, it really is impossible to optimize both at the same time![30] Whatever the case may be, in such serious instances as these I see it as imperative not to leave it to individual developers to decide which standard of fairness should apply. What we need instead is a public, transparently made decision in which everyone has had their say.

Yet the fact that irreconcilable fairness metrics exist also means that selecting a fairness metric and optimizing an algorithmic decision-making system will always put one group at a disadvantage when seen from the perspective of the other fairness metric. Always. There is no solution for treating all groups fairly in every aspect. *Nor is this something unique to digital decision-making.* It applies to every decision involving subgroups that demonstrate different forms of behavior—*even when humans are doing the deciding.*

So what does all this mean for human decision-makers? The issue of how to deal with different groups of people must have come up in the course of their training. Do they hold everyone to the same standards? Take, for example, the question of whether someone is invited for a job interview or not. Does one group of people have to be more convincing than another to receive an invitation? If the same standard of measurement is used but one group of people is known from experience to be less successful, should the error rate among those invited be the same for every group? Or should every group show the same false positive rate? While this topic may well never come up during professional training, this kind of decision is made every day.

With the fairness metric, we bid a final farewell to the backstage area. It is an operationalization, a way of quantifying the social concept of improper discrimination. As with all other operationalizations, the judgements and decisions made about important cases must be determined to be legitimate by society at large and communicated transparently. Yet for the first time, it's now become clear that the various perspectives people bring to the question of distribution are fundamentally irreconcilable in their mathematical expression.

THE ALGOSCOPE

This explains further why the algoscope focuses on ADM systems that directly affect either people or their participation in society. At first glance, it isn't obvious why only this subset of decision-making systems should stand in need of technical regulation. The pulleys and levers used backstage are ultimately the same, after all, no matter what data is being processed.

The rationale lies for one in the complicated legal status of personal data with respect to data privacy rules, but also in the significantly larger number of modeling and operationalization decisions that have to be made in such cases. The more decisions there are, the more difficult it is to test for the OMA principle—whether or not the algorithm and model, with all the modeling and operationalization choices that have gone into them, are in fact capable of granting insight into the question at hand. Yet disregarding the OMA principle raises the likelihood of causing harm to individuals and society at large.

A quick review of the long chain of responsibilities can show us why this sort of situation involves so many more decisions of this sort, as well as a greater number of transparency and accountability requirements than, say, an AI system used on a production line.

*Data:* Data, especially when it comes to data about people, is often *biased*, *incomplete*, and *error-prone*. One part of the data must be *operationalized* (made measurable) in order to get a digital handle on social concepts. Yet an operationalization will always be conceived from a specific cultural perspective and thus represent only *one* potential facet of the social concept. By contrast, when it's an object on a production line that is meant to be evaluated, the input data are likely complete and there's no need to operationalize social concepts.

*Methods:* Most sets of ethical guidelines prescribe that humans must fully be able to understand the decisions made about them.[31] This requirement greatly reduces the number and kind of methods available; in particular, the truly powerful forms of AI that exist to date (neural networks, for example) are not very transparent. Other statistical models such as decision trees may be easier to follow but are often less accurate for it.

*Quality metric:* Where purely commercial operations are concerned—optimizing the way a product is manufactured, say—the quality metric is often clear from the get-go. It usually involves the profit margin or a product property that is easily tested for. For example, the former CEO at Volkswagen, Martin Winterkorn, was famous for using the symmetry and width of the gaps between various car parts as the lone yard stick for measuring the quality of the entire vehicle. If the same width existed all around a part, all was fine. If the two parts weren't parallel, and here and there a gap was wider, all was lost. It gets trickier to settle on a quality metric when people are involved. The effects of false-positive and false-negative decisions are often complex and can't be summarized in a single number. That's why important decisions require social consensus about which quality metric should be observed during training and which alongside it; fairness metrics aren't necessarily used in training itself, for example, but might be important for evaluation when the system actually comes into wider use.

*Fairness metric:* The aspects of fairness and justice only enter the picture for decisions that involve humans or access to social resources; it isn't necessary to consider them when deciding about objects. So long as the objects don't directly impact people, it's unnecessary to make modeling or operationalization decisions.

*Interpretation:* Most of the decisions an algorithmic system will make about a person concern her current and/or future behavior. Does this person have potential? Will she pay back her debt? Is she a terrorist? In most cases, foolproof decision rules don't exist. That means a machine's decisions will only ever be of a *statistical nature*; the risk an individual poses for behaving in a certain way will be given as the share of people similar to her who behaved in a similar way. This makes it notoriously difficult to interpret risk predictions for human behavior. If a machine says that someone stands a 70 percent chance of committing another crime, it doesn't mean they are also responsible for 70 percent of the social cost of a crime and must thus serve 70 percent of a prison sentence. A person either has committed or crime or hasn't; there is no such thing as committing 70 percent of a theft or murder. Such a result is statistical, akin to saying, "70 percent of the people like you have committed crimes." The algorithm determines which group the person supposedly resembles.

> Proceeding by a sort of *algorithmic guilt by association*, decision-making systems thus shift an individual's rate of risk onto the group. This creates a form of *algorithmically legitimated prejudice*.

It isn't always easy to discern the types of decisions such a statistical statement empowers us to make. As psychologist Gerd Gigerenzer and others have repeatedly pointed out, humans are bad at interpreting statistics well.[32]

*Action:* It isn't intrinsically problematic that algorithms and heuristics uncover forms of discrimination in data; what matters is how it is dealt with subsequently. Continued (and naive) reliance on correlations may lead to those forms of discrimination being reinforced, to a preservation of the status quo, or even to balancing out historical forms of discrimination. Qualms like these don't exist with decisions that don't involve people.

*Feedback:* It's only when people are involved that predictions and the resulting decisions can potentially influence the behavior of the affected party. If a person who would not have become a recidivist otherwise receives a sentence based on a false prediction and her life takes a turn for the worse, she may become one now. This in turn can generate one-sided feedback, as it isn't possible to recognize that the person was in fact falsely categorized in the high-risk group. One-sided feedback doesn't allow for the statistical model to undergo dynamic improvement. The phenomenon of one-sided feedback can also be observed in selecting job applicants—applicants who aren't invited then aren't able to demonstrate their ability—as well as with access to education, where students who are turned down aren't able to show that they could have made the grade, so to speak.

And why does the algoscope focus only on algorithmic decision-making systems that learn? AI also includes knowledge systems that store facts, as well as expert systems. Such expert systems include decision trees assembled by people that result in unequivocal decisions. When AI is based on decision rules that have been put together by people and stored in expert systems, many of the problems mentioned here don't weigh as heavily. Both the quality and fairness metrics have received sufficient attention from experts for clear decision rules to be formulated. Furthermore, this kind of rule system is usually understandable to the layperson. There's zero wiggle room for differing interpretations, and any potential actions are fixed. In most cases, the rules have either already been fed through an involved feedback loop or were so clear already that it wasn't necessary. Here too, using AI with a learning component is only plan B: if an expert system can be built, it should be. Nevertheless, in the future we will come increasingly to rely on machine learning as it is the only thing capable of handling the truly interesting situations.

> This explains why the algoscope is right to focus its attention on ADM systems based on statistical models trained from data that impact people or access to social resources, and thus social participation itself. Whether the systems do this automatically or merely in a "supportive" role is of lesser importance.

Finally, the preceding discussion has shown that the focal point of any inspection does not lie with the specific algorithm or heuristic that has been used. That is why the term *algorithmic decision-making system* has begun to appear more and more frequently throughout the past couple of chapters. Such a "system" is made up of the data, the method of creating the statistical model, all modeling and operationalizing decisions, the statistical model itself, *and* the algorithm that arrives at a decision, aided by the statistical model.

And how might we go about developing one of these systems so that they make ethical decisions that also make sense for society? That's what we'll talk about in the next chapter.

# HOW TO STAY IN CONTROL

Sometimes algorithms really have no idea what to do with me. When I first began my doctorate in computer science, I received a striking number of ads suggesting easy, painless ways to increase the length of my . . . my what? Sorry, you won't find that body part here. I suspect it was the large number of programming questions I was entering into search engines at the time that led the algorithms distributing the ads to think I was male. Now that's a wrong decision on the part of the algorithm, no doubt about it— but does that also mean we immediately have to trot out the full arsenal of every possible means for control or regulation? What options do we have, anyway? And who's supposed to decide when that happens or the extent to which a decision-making system stands in need of technical regulation— and by which criteria?

Computer science research draws a distinction in this regard between *transparency requirements* and *accountability requirements*. Transparency requirements ask for information about all the many decisions that went into developing an ADM system, such as the method used, the quality metric, or how well the system did on a certain test set. They encompass every piece of information a company could give on the development, testing, and use of an ADM system. Accountability requirements, on the other hand, ask for even more in order to perform the necessary analyses themselves. That includes access to parts of the ADM system itself, such as the training data, the system, or its code. One might require that experts check the training data for bias, for instance, or retrieve the system's decisions on a made-up set of input data to look for unjustified discriminatory behavior.

In essence, different transparency requirements must be met for society, for the people affected, and potentially also for groups of accredited experts. They might include transparency in terms of the quality and

fairness metrics, the type of input data, and the method of machine learning used, as well as the evaluation process and its results.

As I mentioned earlier when discussing possible forms of discrimination
in ADM systems, black box testing can check for any number of issues if
it's an option. To test a company's software, for example, you could create
hundreds of slightly varied job applications, then send them on to the company. One possible accountability requirement in this case would be for the
software to be open enough to allow for black box testing. This isn't to say
it would be necessary for anyone in theory to be able perform such a test;
opening the software up to such an extent could easily lead to the system
being analyzed and potentially manipulated by third parties.

Still, there a few instances in which black box testing isn't enough on
its own. In more serious cases, greater insight into the workings of the
system—the data basis and the trained statistical model, for example—
may be necessary. That includes the requirement that the statistical model
show *accountability*, in computer science speak. This last point may sound
harmless, but in fact it severely restricts the number of methods available
to choose from to a small subset of weak methods that are not as good at
finding patterns. Finally, there are the ideas that must be banned flat out.
And just how are we supposed to know what goes where?

## THE RISK MATRIX

Over the last several years, I've developed a regulatory model consisting
of five levels. The levels are divided first according to a decision-making
system's potential for causing harm, and second by the options that exist
for calling the system's decisions into question and possibly changing them.
The potential for harm includes damage to individuals on the one hand,
especially in the cases of *wrong* decisions, and the potentially much greater
harm to society that the decision-making system may inflict on the other.
It will soon become clearer exactly what I mean by this.

In some cases, both forms of harm are negligible. If one of the many
online marketplaces out there tries to sell me a XXXXL shirt in plaid, I can
simply switch platforms: Who wants plaid, after all? Things look a little
different, of course, if my job application is rejected! In that case, the harm
inflicted by an (imperfectly) functioning system for evaluating job applications largely comes down to individual instances of wrong judgment.

Those decisions harm the people who have been wrongly turned down—and in turn the state, which now has to provide for them, or perhaps the company, which now mistakenly gets a less talented or suitable employee. Especially when used in an all-encompassing manner, this kind of system must be examined for systematic discrimination. No part of the damage caused to the state or the employer is superlinear, however, meaning it will not exceed the sum total of harm caused to individuals.

The next two examples are another can of worms. A newsfeed or YouTube page that suggests content I have no need for may not cause me much harm as an individual user. If the wrongly distributed content contains conspiracy theories, on the other hand, this may cause a great deal of harm to society as a whole. In this case, the damage to the individual—a couple of minutes spent on unsuitable material—is rather limited, but the overall harm to society can be tremendous. One current example is the debate surrounding vaccines, which is fueled by seemingly scientific material that casts doubt on vaccines despite their proven efficacy. On the one hand, the level of harm increases through the potential consequences of measles for the larger group of kids who now aren't vaccinated against the disease; that's the linear part. At the same time, it endangers people whose fragile state of health won't allow them to be vaccinated—one aspect of superlinear harm. On top of all this comes a diminished trust in science overall, a type of harm that also isn't linear. Rather, it can so heavily upset the delicate balance between establishing facts and scrutinizing those facts that scientifically tested knowledge comes to be regarded as just one out of many "opinions."

As a second example, surveillance software holds even greater potential for inflicting harm. On the one hand, the software stands to cause a great deal of harm to individuals who are wrongly placed under suspicion as false positives. It also causes harm to society by not detecting criminals (false negatives). Yet it is society itself that is truly weakened when it is brought under constant surveillance, a condition that impinges on fundamental democratic rights. In this case, the sum total of harm caused to individuals is large, but so is the overall damage caused to society.

Table 1 summarizes the four examples from this section.

Analyzing the potential for harm in this way remains by necessity approximate and always takes the form of a *worst-case* scenario. Whatever that case may be, though, the potential for harm will always depend on the concrete role the decision-making system is playing in a given social

Table 1
Examples of algorithmic systems whose decisions cause lesser or greater harm to individuals and society as a whole

|  |  | Harm caused to individuals | |
|  |  | Small | Large |
| --- | --- | --- | --- |
| Harm caused to society as a whole | Small | Wrongly evaluating a shopper's fashion sense | Wrongly evaluating a job application |
|  | Large | Spreading conspiracy theories through social media | Surveillance software in public areas |

process. Using the predicted risk of recidivism to assign therapy *after* prison time has a different potential for harm than if that same prediction is used by a judge *before* a sentence is handed down to send the person to prison.

The potential for harm, then, is one dimension that determines how transparent a decision-making system must be and how tightly regulated. A second important aspect considers the extent to which a person is able to work around the decision, get a second opinion, or switch over to another system. In the broadest sense, it addresses the issue of monopoly, as well as the number of oversight measures and opportunities that exist either to contradict or reevaluate a decision. A decision-making system applied in a monopolistic setting—for example, by the state—must be monitored more closely than software that exists on the market among many other options. Yet even for systems that do hold a monopoly, the greater the number of nondigital opportunities that exist for establishing oversight and voicing objections, the more transparently these are structured and the more public they are, the shorter the list of regulation requirements. In brief, this second aspect gauges how easy it is to discover, contest, and change wrong decisions.

In the risk matrix in figure 38, the harm a decision-making system stands to cause is shown increasing from left to right, while its contestability/monopoly moves from top (large market, numerous chances to contest results) to bottom (small market, few chances to contest results). This puts systems that require little to no regulation at the top left and those requiring strict regulation at bottom right.

It's important to note that aside from their potential to cause harm, there are obviously *benefits* to using algorithmic decision-making systems as well.

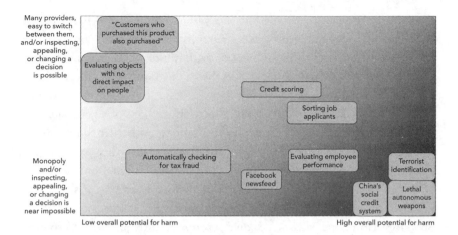

Figure 38
A risk matrix for assessing the degree of technical regulation necessary for a sample arrangement of algorithmic decision-making systems.

While those benefits are *irrelevant* at first when considering the degree of technical regulation required, analyzing them can help us decide whether or not a decision-making system makes up for the trouble and expense involved in its technical regulation. Diagnosis software used in every hospital and doctor's office across the US would be monopolistic and be assigned a large potential for harm, for example, but the potential overall benefit would likely be so great that society would be willing to accept whatever technical controls were necessary for it.

Three trends become visible from the risk matrix:

1. So long as no other alternatives exist, state use of an algorithmic decision-making system gives it a stronger aspect of monopoly, moving it lower on the matrix. The more opportunities there are to appeal a software's decisions and the more people available to speak with about those decisions, the less monopolistic it is. Analog opportunities to contest decisions, then, can improve a system's position on the risk matrix.
2. Personalized services that show each user a different result generally cost more to test with black box analysis and often show a greater potential for causing damage, moving them further to the right in the matrix.
3. The greater the number of people a system affects, the greater its overall potential for harming individuals and usually society as a whole.

If you take the two dimensions we've been discussing—harm potential and the opportunities present for identifying, disputing, and changing wrong decisions—you can arrange algorithmic decision-making systems accordingly in a 2D matrix. Figure 38 shows one possible configuration.

The examples given here for more or less unproblematic decision-making systems include one that recommends products of a general nature and another that makes decisions about products—whether to take defective screws off the production line, for instance. Another example is a system already used in a number of federal states throughout Germany, which automatically adjusts taxes for annual income as a way of shortening processing time. In this case, the potential damage is low, but not zero. Scoundrels might get their hands on the software, after all, then try to mask income in their tax forms without the algorithm noticing from the comfort of their own home. This is in fact quite a common problem with algorithmic decision-making systems: whereas before large numbers of people were making these decisions and it was impossible to spy on all of them, there's now a great deal of incentive to ferret out the weakness of a single opponent (the software). Still, despite the state monopoly, there aren't too many problems in this case, thanks to tax advisors and the large number of points at which decisions can be contested.

The algorithm Facebook uses for its newsfeed gets a significantly worse rating as the system responsible for selecting and deciding the order in which posts by friends or companies appear on a user's start page. The scandals over the past few years have made it painfully clear just how much havoc third parties can wreak by manipulating delivery of fake news and conspiracy theories. What's worse, users have practically zero influence over Facebook, and we computer scientists aren't able to measure the extent of filter bubbles or echo chambers.[1] So long as that remains the case, there's a high amount of potential harm—that is, the damage we can't rule out as impossible—and only limited chance for redress.

Credit-scoring systems receive roughly the same rating as Facebook's newsfeed algorithm in the risk matrix, although in this case a broader market exists, composed of different banks. Provided there are enough evaluation systems within the field that are distinct from one another, the individual systems rank higher than Facebook's algorithm.

The two systems for evaluating employees are also worth mentioning. So long as it is only applicants that are concerned, the market offers multiple systems (job platforms) for their use. If, however, a company is performing an internal evaluation of its workers, then an individual employee would be hard-pressed to find a second opinion. If there are no expanded mechanisms for transparency or oversight, the system slips farther down in the risk matrix.

Finally, the matrix shows three systems in the far bottom-right corner, all there for different reasons:

1. *Lethal autonomous weapons*, as they are called, inflict the greatest possible harm to individuals when they make a wrong decision, killing a person they're reasonably certain they have identified. In a wartime situation, it's neither possible for the condemned person to appeal the decision nor possible for them to get any insight into the machine's decision. The overall effect on society is also a hot topic among jurists. Facial recognition that is 100 percent accurate does not exist now, nor will it ever for the technical reasons discussed previously—which means that machines must fire when their results are "sufficiently certain." In certain situations, this dispenses with the presumption of innocence, a legal principle that clearly shouldn't be abandoned unless absolutely unavoidable.

2. Software for identifying terrorists lands in the bottom-right corner for a different reason. The truly powerful methods of machine learning are all data-hungry. The number of legally convicted terrorists is small. So long as you suppose that different terrorist groups attract different people and that the respective cultural milieu plays a part, you would need to train systems by terrorist group and by country, further reducing the number of data points available for each individual case. This suggests that the methods of machine learning should only be used in this context to get a better idea of what's going on in an area of interest (data mining), but not to make decisions directly. For technical reasons, then, so long as there are a limited number of data points, algorithmic decision-making systems aren't well-suited for identifying terrorists.

3. Finally, there is the social credit system used in China, which, as I've discussed, is conceived of as a way of providing citizens with constant feedback about their behavior and rewarding them when it is deemed "good." There are real-world consequences to this; citizens with too low of a score have been denied access to high-speed trains, for example. The amount of surveillance this requires gives cause for concern, as does the

power gap between citizen and state, which only increases with the software's use. Having ties to "bad citizens" can also lower one's own score, effectively guaranteeing that government opposition will be isolated, socially speaking. Thus, from a Western democratic perspective, the potential for harm caused by the system is great enough to put it in the lower-right corner of the matrix.

THE FIVE REGULATORY CLASSES

When pursuing research at the University of Kaiserslautern's Algorithm Accountability Lab, we categorize algorithmic decision-making systems according to one of five classes, based roughly on where they appear in the risk matrix (see figure 39). These classes in turn determine what sort of transparency and accountability requirements should be put in place.

*Class 0:* The systems in this class show such low potential for causing harm that no technical regulation seems necessary at present. In case of doubt, we might take a second look to test whether discrimination is an issue, for example, or the system causes any other forms of harm. If the suspicions prove justified, the system's potential for harm is raised and it automatically enters a higher class with stricter requirements.

*Class 1:* The algorithmic decision-making systems in this class show a nontrivial potential for inflicting harm and must be continually monitored. Accordingly, the system must have an interface that allows for this type of analysis (an accountability requirement). In addition, society should be informed as to which quality metric was used to train the system, as well as the method(s) of machine learning employed. Last, but not least, it's important to understand the system's role in whatever social process it is involved in. Is it playing a supportive or auxiliary role? Is the decision automatic? What consequences will these decisions have, and what opportunities for contesting them exist? This list of questions constitutes a set of transparency requirements.

*Class 2:* In this class, there is an increased potential for harm, and few opportunities for challenging a decision exist. In addition to the requirements listed for Class 1, in Class 2 it also becomes necessary to get a more precise understanding of the kind of input data used (a transparency requirement). Society should be allowed to inspect the system's quality assessment for itself (accountability requirement).

*Class 3:* Class 3 systems are seen as having the potential to cause a great deal of harm. Direct access to the mechanisms behind the decision is a must in order to get an idea of which properties are leading to which decisions. This means that algorithmic decision-making systems in this category must be trained using machine learning methods that allow insight into the decision rules they discover. Today, only very few methods are considered to fall into this class, among them decision trees (see chapter 5); logistics regressions, as in the example of the Austrian Public Employment Service (see chapter 10); and a few others. This represents a severe restriction, because these methods are in general much less effective at uncovering patterns in data. We stipulate further that input data for systems in this category must be accessible—for example, in order to inspect for forms of discrimination already present within the data.

*Class 4:* This class contains decision-making systems that shouldn't exist in the first place, either because they show grave potential for harm or because they can't be implemented for technical or legal reasons.

Figure 40 presents a summarized version of requirements for transparency and accountability, originally presented in greater detail in a study for the Federation of German Consumer Organizations.[2]

Incidentally, it must be decided on a case-by-case basis who carries responsibility for ensuring these requirements are met. In most cases, that

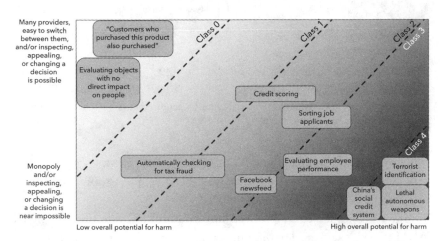

Figure 39
The risk matrix is divided into five classes, each with their own transparency and accountability requirements.

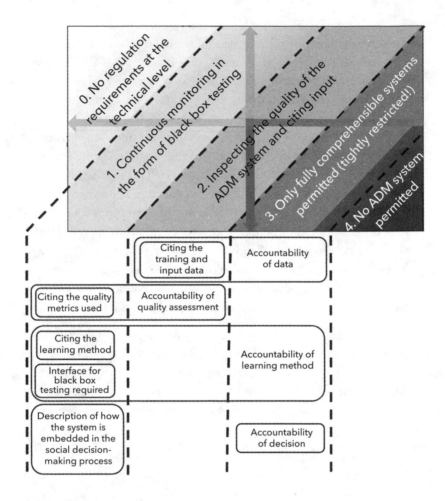

Figure 40
A summary of the transparency and accountability requirements for the five classes of
algorithmic decision-making systems.

will be the general public, though in the area of sensitive military software
it would be conceivable to limit transparency to democratically legitimated
and independent institutions.

The regulatory model I propose here can serve as a first filter before an
ADM system is subjected to the rules of an algorithm safety administration.
The important step here is that, in contrast to the original idea of Mayer-
Schöneberger and Cukier, it would select those systems whose usage might

cause harm, not just any algorithm independently of the context it is being applied in. The five regulatory classes within the model reflect years of thinking about how and where errors can occur in the design of algorithmic decision-making systems.[3] While this model has already proven its mettle in many rounds of discussion, it remains a long way off from being directly applicable and is doubtless in need of further refinement. Still, it is the first of any model I'm aware of to differentiate between regulation requirements for decision-making systems. This is an important way of allowing for innovation so that not every idea gets bogged down in requirements from the beginning. The model features a second property that allows for innovation: a system with few users in the beginning poses a much smaller risk of inflicting harm due to the limited number of people it impacts and can therefore be tested out in Class 0. Should the number of people affected by the system grow, reclassification may be in order. To further minimize legal requirements, those implementing a system can help ensure its potential for harm is given real-world analysis: the more transparent a company voluntarily makes its software, the easier it will be to identify the places where problems may in fact arise.

The regulatory model is silent, by the way, about who should make the call as to which class a system is assigned and when that happens. Let's take a look at where we might find experts who could tackle these kinds of questions.

## WHY THE WORLD NEEDS SOCIOINFORMATICS

There's only one way of assessing the overall harm an ADM system is capable of causing, and that is by understanding how digital systems *interact* with individual people, institutions, and society at large. Such a set of relationships makes up what my discipline, socioinformatics, would term a *sociotechnical system*. This entails a shift in perspective for computer science. To date, software design has been defined more than anything by *requirements*, in the sense of functionalities demanded by whoever has been identified as a stakeholder. A *stakeholder* in turn is anyone perceived by the software designer as holding a legitimate interest in the software's design. Today, however, a stakeholder is defined rather narrowly. Let's take completing search terms as an example. In the classical sense, the stakeholders would be the company and the search engine user. Consulting these two groups might generate the requirements that suggested search term completions must appear within a tenth of a second, and there can be no more than seven as the function won't be used otherwise. The company might add a requirement that on the whole the software must contribute to more customers using its search engine.

A traditional software engineer would then consider what kind of search term completions might save the user time. "Maybe this user will be searching for what lots of other people have already searched for today," she might think to herself. If a person then clicks on that suggestion while entering a search term, it's rated a success. This allows the system to be improved upon dynamically; when the next person searching for something similar comes along, the term will appear near the top of the list. A form of *positive feedback* thus emerges; whatever many people have searched for in the past will be offered to many others as a search term.

In the beginning, Google's method of search term completion functioned pretty much in the way just described—until the first scandals showed up, that is. As it turned out, Google's model didn't account for the fact that clicking on a suggested search term didn't necessarily mean the user had wanted to enter that specific term in the first place. I got a taste of this myself in the summer of 2015, when I went to look up something about Chancellor Angela Merkel. I don't recall anymore what I was searching for, but I do recall exactly what Google suggested as a search term: "Merkel . . . pregnant." Merkel? Pregnant? It struck me as somewhat doubtful, seeing as how she was over sixty. My curiosity got the better of me, though, as it did many others it seems. "Merkel pregnant" was one of the more popular search terms at the time, something that Google Trends—which you already know about from chapter 4 on big data—can confirm (see figure 41).

I had fallen for one of German satirist Jan Böhmermann's jokes, which the search term completion algorithm had then amplified far beyond any actual interest in the search term itself. Even if I no longer know why I typed in the term "Merkel" that day, it certainly wasn't because I was looking for information about the (im)possibility of the chancellor being pregnant. But the algorithm tempted me, and by clicking on the suggestion, I sent the message that it was exactly what I had been looking for! The second peak in April 2018, by the by, comes from a similar April Fools' Day joke that was then disproportionately disseminated by the algorithm. Incidents like these are certainly uncomfortable enough for those affected,

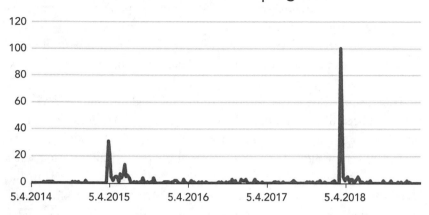

Figure 41
Relative level of interest in the search term "Merkel pregnant" over the past five years.

and I had to think long and hard about whether to raise this issue here in the first place. There are other cases where the same algorithm has had even more unsavory effects. One hotel owner had to wrestle with Google's suggestion to anyone searching for their hotel that the user may in fact be interested in the murder that had been committed there years earlier; search term completions have insinuated that some women had backgrounds in the sex industry; and politicians are regularly associated with scandals with which it has been proven in court they had nothing to do.

What's going on here? Well, for starters, there's the role that the algorithm is playing in the social process of finding information. When non-emotional subjects like making sense of a computer error are the focus, the algorithm functions marvelously. Those searching for "Windows crash" will be grateful for the suggested terms, which already provide hints as to the problem that is most likely behind the crash in the moment. By clicking on the appropriate suggestion, the user sends an authentic signal that it is precisely this search term in which she's interested. What matters here is that the person was already interested in the question, even if they didn't know exactly how to formulate it before they saw suggested searches.

If, however, the search completion picks up on a signal that is more akin to clickbait, such as a potentially juicy scandal that has everyone scrambling to find out more, we enter the realm of self-fulfilling prophecy. Something that nobody was searching for becomes the most popular search term simply because it was suggested. Only in terms of an overall system are we able to see that in certain social constellations, an algorithm that basically runs without flaw can become a vector for malicious gossip. In this case, where there is smoke, there may well be no fire at all.

This type of issue might have been avoidable had the search term's relevance not been directly modeled by the number of clicks. Once again, it's a question of getting the *operationalization* right!

Today, search suggestions may still lead people to "information" that they would never have searched for otherwise. When I looked up the search term "vaccinations are . . ." in April 2019 (see figure 42), one of the suggested term completions was "blasphemous" (*gotteslästerung*). Under the search term "vaccinations are harmful," you can find both videos and facts on the subject of vaccinations provided by reputable news sources and health institutes and sites run by those who oppose the practice.

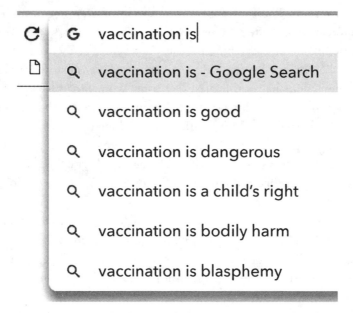

Figure 42
Search terms suggested by Google for an inquiry beginning with "vaccination is . . ."

This example shows once again that we can often only begin to get the measure of a given piece of software's overall potential for causing harm by taking the entire process into account. Individual and group psychological motivation play an important role in this context and need to be weighed against the incentives a particular software offers; the software and its effects also have to be considered from legal, economic, and ethical perspectives.

This is why we train students studying socioinformatics at TU Kaiserslautern to be software engineers who never lose sight of the big picture. Students receive a basic education in economics, law, ethics, sociology, and psychology to give them an appreciation for humans' varied interactions with one another. They draw on a number of models in their attempts to replicate and, when possible, predict the way different social actors connect with each other, including game theory, network theory, and even statistical physics. Analyzing and responding to the societal impact of software that connects the data, actions, and behavior of millions may still be in its infancy, methodologically speaking, but to my mind the systematic perspective it brings is absolutely essential. Having a team member trained in socioinformatics, then, seems like a bright idea wherever software with potentially large social side effects is involved. He or she won't have an answer for every question, but will know when to call in which expert. That clears up one part of the question about who might be able to analyze AI's impact on society: experts trained in the field of socioinformatics. Yet they will form only one part of an interdisciplinary team, of lawyers and economists, maybe also psychologists, sociologists, and philosophers. Before I move on to answer the second question about when that should happen, I'd like to step back for a moment. Why would we possibly want machines making judgments about people in the first place, anyway? The answer has to do with different ways of thinking about people and how they make decisions themselves.

# WHO WANTS MACHINES MAKING DECISIONS
# ABOUT PEOPLE, ANYWAY?

The whole notion of using algorithmic decision-making systems doesn't just come from their development. In most cases, they're meant to improve on an ongoing social process, either by supporting or replacing people in their decision-making. Examples include recidivism prediction systems in the US or the broad availability of systems for evaluating job applications, but also weapons systems that fire automatically once they believe they've identified the intended target.

Sometimes social processes rely on ADM systems to be possible in the first place—for example, when the sheer number of decisions that have to be made means that humans can no longer reasonably be expected to manage. That applies for all the recommendation systems that operate in the digital realm, but also those for rating citizen behavior, as with China's social credit system, which would be inconceivable without big data. In these cases, machine decisions seem like the only viable option.

In this chapter, I'd like to focus on decision-making systems that are intended to support or replace human decision-makers in established social processes like reaching legal judgments, the job application process or identifying talent within a company, education, and obtaining credit. When the goal isn't cost efficiency first and foremost but a desire to make a given process "more objective," "less discriminatory," or "better," we need to keep in mind that there are different images of people bound up in such a wish.

## THE IMAGE OF PEOPLE UNDERLYING THE USE OF ADM SYSTEMS

In most cases, such a desire rests on one of two images of people, both of which make assumptions that merit close inspection. The first views humans as more or less unskilled in the art of decision-making. This has become such common fare in contemporary debate that you might almost come away with the impression that all people ever do is make bad decisions! Yet in most of the instances I looked at, how bad human decisions actually were hadn't even really been measured before an ADM system was set in place. Yet that is something you would have to determine before you could say whether such a system did in fact subsequently improve the social process. This means that when developing algorithmic decision-making systems, *defining a quality metric isn't enough on its own*. At the very least, a second metric is required: one that measures whether the desired improvement came about. As mentioned before, in 2018 the ACLU pulled an about-face regarding the use of pretrial risk-assessment systems because the systems made no discernable improvement to the overall process.[1]

The second image of people takes even more for granted, contending that a person's future behavior can be derived from either his own behavior in the past or even that of others. Three quite far-reaching assumptions undergird this image, which I discuss using the example of a system for evaluating job applications:

1. The underlying notion is that whether or not an applicant will be a good match for a company is based almost exclusively on their properties as a person: the training they've received and their personality traits (*property-based behavior*).
2. This implies a second assumption: that all essential properties or characteristics responsible for a person's behavior can and will be taken note of (*the possibility of observing and operationalizing all causal parameters*).
3. A final assumption is that an applicant's suitability can be predicted on the basis of people with similar profiles (*transferability*). An applicant has never worked for the company before, after all, and there is only limited information about them, so the algorithm classifies them in a *group*. The groups determined by the algorithm are held under group liability, as all people within it receive the same prediction.

Any of these three assumptions can either be right or wrong independently of the others. I've still yet to come across a single textbook or handbook that names them explicitly, however, or calls for their validity to be inspected before the statistical model for predicting future behavior is put to use. Why might that be important to do?

Remember the decision tree in chapter 5 based on the passenger data from the *Titanic*? Well, imagine now that we built the tree based on the assumption that a passenger's odds of survival essentially depended on their properties or characteristics. The trained tree would then be used as a way of allocating the most support to those with the best chance of surviving similar catastrophic situations with limited resources. Aside from assuming a *property-dependent chance of survival*, and in analogy to the assumptions listed for the job application evaluation system, I would also take for granted that observing the incidence of survival effectively depended only on people's properties (*unimpeded observability of causal parameters*) and that those properties would also play the key role in other shipwrecks (*transferability*).

I'm aware that this thought experiment is grotesque, but it is so largely because most of us would allocate whatever limited resources were available to the people who otherwise wouldn't stand a chance of surviving. It was exactly this sort of thinking that led to the maxim "Women and children first."

The question we're concerned with here, however, is whether these three assumptions hold water in the first place, so to speak, so that we can then draw an analogy to our system for evaluating job applications. It is largely the second assumption that takes a hit in the case of the *Titanic* passengers. The most important variable—the one that was primarily responsible for the behavior "survived"—goes missing here: namely, whether a person received a place aboard a life raft or not. Had this information been included in the data set, it would have explained who survived with just about 100 percent accuracy. The central parameter isn't observable by the machine, in other words, so the statistical model learned from the data gives the properties that tend to lead to a passenger getting a place in a life boat, but not who is more capable of surviving.

It would be appalling if rescue workers in the past had acted accordingly in similar catastrophes, directing whatever scant resources were available exclusively to those a decision tree had ascribed a high chance of surviving. Doing so would only exacerbate unequal treatment! Such a point of view strikes us as so obvious that you're probably shaking your head at the idea of someone crazy enough to actually model surviving the wreck of the *Titanic* as a property-based behavior.

We run into a similar problem, however, when trying to predict recidivism rates for criminals or what an employee is likely to accomplish. In these cases, too, machine learning assumes that whether or not a certain behavior is observable in a person—recidivism and successful employment, respectively—will depend only on her measurable properties. Yet with criminality, doesn't the social environment into which a person is released also have a say in determining whether she commits another crime? Can the social situation—one that is constantly evolving, moreover—be observed and operationalized just as effectively as property parameters (the number of previous crimes and age)?

With this question, we can see that here, too, there's more involved than simply measuring or predicting future behavior. *Should* a person be assigned a higher risk of recidivism if any of the reasons for the risk lie outside that person? If, for example, a child grows up poor, receives a bad education, and experiences violence at home, we have to ask ourselves whether these "properties" can be used as such in the first place. The question is one for legal philosophy: Is it legitimate to assign a person to a higher-risk pool if their parents were also criminals? Is that permissible in some social contexts but not in others? What if it is precisely these properties that dramatically improve the quality of prediction? Should they be used in that case? Are we even obligated to use them?

Once again, we find ourselves faced with the question of what role the ADM system plays in the overall process.

## SHOULD MACHINES BE ALLOWED TO EVALUATE PEOPLE?

Along the long chain of responsibilities, it is *interpreting* a machine's results and choosing an *action* that first determines what kind of social value an ADM system will have. In the previous chapter, I discussed several instances in which discrimination that was already present in the input data had a negative impact on the social situation—systems for evaluating job applications that were poorly equipped to handle image or speech recognition, for example, or to distribute advertisements for available positions. This doesn't operate as a law of nature, however; if algorithms help to uncover unjustified forms of discrimination present within large bodies of data, it's up to society to decide what to do about it.

To give one example, the Austrian Public Employment Service— the *Arbeitsmarktservice* in German, the Austrians call it AMS for short—is currently testing out one such decision-making system. One part of the organization's work as an unemployment office uses an algorithm to sort unemployed people into one of three categories: (a) those with a good chance of quickly making it back on the job market, (b) those with a very poor chance, and (c) everyone else. The limited resources AMS has at its disposal—offering continuing professional education is one example—are largely supposed to benefit those in the third category.

The properties used to sort people include sex, age group, which state the person resides in, their level of education, any health impairments, care-giving duties, occupational class, earlier professional career, and a report on the regional job market.[2] The system was trained with logistic regression, a method whose statistical model is relatively easy to interpret.[3] The heuristic behind the method weighs all the properties presented to it in such a way as to score as high as possible on the quality metric. Once again, accuracy was selected as the quality metric—that is, the proportion of correct decisions out of all decisions made. If the heuristic decides to weigh one property negatively, then many people in the data set with that property will be unlikely to quickly find reemployment. A positive weighting, on the other hand, indicates that a property is helpful.

No matter how you slice it, the results are depressing. Women, immigrants, the elderly, the disabled, and those caring for others all have a harder time finding work. According to the statistical model, a single mother over thirty with two children who's been on parental leave for the past three years will be much more hard-pressed to find a job than a man under thirty with health problems. Is such a result already discriminatory? Should an ADM system not be shown these properties in the first place if it will then discriminate according to them?

Well, for starters, the finding represents an analysis of the current job market, which means the consequences of the categorization are what truly matter. And since the algorithm sorts a higher share of women in the third category, they will tend to receive greater support for continuing education. This is what has led Johannes Kopf, the chairman of the board at AMS, to argue that using the algorithm will ultimately even out social inequality.[4]

Seen purely in terms of its functionality, the algorithm's high potential for inflicting harm paired with the broad monopoly of AMS would definitely make it a Class 3 on the risk matrix outlined in the previous chapter. An unemployed person could of course try to find work another way or organize their own professional training, but overall the organization is just about the only game in town. The algorithm also threatens to inflict a great deal of harm by systematically strengthening discrimination. The risk matrix would thus mandate strict transparency and comprehensibility requirements for such a system.

This explains my tremendous skepticism when I read the initial reports about the algorithm. I was pleasantly surprised to find that the team of developers behind it, led by Professor Michael Wagner-Pinter, had voluntarily met nearly all of the transparency and accountability requirements given earlier by releasing a study that explained the algorithm in a generally comprehensible way.[5] I still had my doubts, though, and asked for a phone call to better understand the philosophy behind the system's development.

## A BLUEPRINT FOR TRANSPARENCY

On our call, the ease and deftness with which Wagner-Pinter described the technology behind the classification system gave me an inkling of just how many times he must have done it over the past few months. That, and how rarely the people he spoke to must have even opened the report, in which stood written all the particulars he was now discussing over the phone! With me it was the exact opposite: I had combed through the report at least three times to understand precisely what had been done, and my questions concerned a couple minor points that were still unclear after a third reading. The report itself contained information about the selection and different kinds of input data, as well as the method of machine learning chosen.[6] It also explored how good the system's predictions were, further breaking them down for both sexes. The report really is written in such a way that a layperson could get an idea of how the system functions, while an expert could more or less follow how the system was constructed. What was more, it gave the formula for one of the two statistical models used and explained its impact. Nevertheless, one coefficient in the logistic regression still puzzled me: according to the formula, the machine took frequent visits to the employment office over the past four years as a sign that the person would find work again quickly. From where I sat in Germany, at least, such a finding struck me as rather unintuitive.

I had a hunch, though I hardly dared believe it might be right. "Professor, I've noticed the statistical model weights frequent visits to the employment office quite positively—is it possible that seasonal workers are mainly responsible for shaping the trend?" Wagner-Pittner laughed, then confirmed my suspicions. Austria is famous throughout Europe for its lovely tourist destinations, exciting ski options, and wonderful food; on average, it receives more than thirty million tourists annually. This explains why many of the jobs in the country are seasonal. "Did you know that even tractor manufacturing has a season, besides agriculture and tourism?" he quipped in a velvety Austrian accent. No, I hadn't known that, actually: case closed! Truly difficult cases can be spotted when a person takes other measures in addition to frequent visits to the employment office. By the end of the phone call, it turned out I didn't have any outstanding questions for the time being, so far as the statistical model was concerned.

Something else Wagner-Pinter had to say did leave me astonished, however: "With classification, our goal was to create individualized representations of AMS's clients." Really? With just twenty-two variables? One of which is occupational group, putting the manager next to the stable boy in one category and the teacher next to the hairdresser in another? I shook my head impatiently—a bad habit of mine when I don't agree with what someone's said. Fortunately, Wagner-Pinter couldn't look through the telephone to see what was definitely a premature reaction on my part. "Before," he went on, "employees working at AMS mostly looked at how long a person had been unemployed and naturally didn't make any predictions as to which properties would help someone find work quickly."

This is actually an important point to consider, all misgivings about machines deciding for humans aside: a machine can still improve a social process even if it doesn't work perfectly. The point here is to compare it to how humans have made those decisions in the past. In theory, the AMS algorithm can distinguish among eighty-one thousand groups, around eight thousand of which appear in sufficiently high numbers. That means the computer can differentiate much more effectively than humans.[7] While the algorithm will thus create groups defined by the algorithm alone, they will still be more highly differentiated and nuanced than what AMS employees had managed in the past.

On the whole, I have to say the report on AMS' algorithm verges on the ideal, very nearly serving as a blueprint for ADM systems when they are used by the state. The input data are described; the methods of machine learning are sufficiently transparent, even if not all the details are present; and the quality of the model is measured—though I might have wished for a little more information here than just the rate of accuracy. We're still talking about a Class 3 algorithm on the risk matrix after all, which to my mind makes it necessary for independent experts to evaluate statements on the system's quality. Some form of external comprehensibility is still lacking, in other words.

Yet Wagner-Pinter had something that might even top that: his company's rules on social sustainability. "Would you be interested in seeing them?" he asked. "Absolutely," came the reply—and I meant it.

RULES ON SOCIAL SUSTAINABILITY FOR EMBEDDING ADM SYSTEMS

Wagner-Pinter doesn't take every request he receives to develop decision-making systems. The client has to convince him that the algorithm will only ever be used as a second opinion on a decision made by a human and that the result will be discussed between the person making the decision and the one affected by it. It's important to him that the model views people in a sufficiently differentiated light, which in turn depends on large quantities of training data. What's more, the model's lifespan has to be attuned to changing realities. For the AMS algorithm, that means that it has to be recalculated every twelve months based on the most recent data. Last, but not least, it's essential for Wagner-Pinter that the systems are able to forget. "We only look back four years into the past. People have to have the right to continue developing!" he adds.

There's a pressing need for broader public discussion about these and other self-imposed rules, which *to my mind represent an important baseline for taking a values-led approach to implementing artificial intelligence*.[8]

One important rule is missing here, however, one that isn't actually the responsibility of the developers but the client. That's checking for whether use of the algorithm does in fact lead to unemployed people receiving help "more efficiently"—a real-world test over the long term. As mentioned earlier, at present a process is missing that would allow for at least a handful of select institutions and external experts to conduct tests of their own on the system.

This brings me back to the question of whether people should be evaluated for their risk of recidivism based on properties that they can't change, such as their childhood. The answer? It depends on the context in which the algorithm will be used. Allocating job trainings as the AMS algorithm does, for example, should without question be evaluated differently than typesetting a sentence, as Knuth and Plass's algorithm does.

## AT A GLANCE: HUMANS VERSUS MACHINES

I myself am still quite skeptical as to how often we can truly expect machine learning algorithms to offer people meaningful support or even replace them in their decisions about others. I find the images of humans underpinning such methods quite distorted. Of course, humans sometimes behave irrationally when measured according to the grotesque oversimplifications of the *Homo economicus* model. Yet when measured against the constraints and side conditions under which humans have to make decisions, we do act rationally. As humans, we are limited in our ability to absorb data and must conserve energy, which means we can only make so many decisions per day. We don't consider every single option when we're looking to buy a car, for example. Instead, we decide on one or two possible car manufacturers early on, then narrow our search from there. In many situations, reducing one's possibilities in this way is less than ideal. In this instance, relying on a machine that can search through nearly unlimited amounts of data for any number of correlations and draw meaningful details even from the most minor of connections isn't a bad idea.

What continues to baffle me, though, is why machines should be allowed to develop hypotheses from observations, then use them untested to judge other situations—something that every laboratory the world over would consider unscientific. Especially in situations where people's lives could be severely impacted by the decision reached, it shouldn't be possible without first testing the system exhaustively. Nor is it enough to test the system on its own in a vacuum. Rather, it has to be considered in terms of its position in the social process it is supposed to improve.

It's just as perplexing to the natural scientist in me why each of the individual correlations found by machine learning methods are not tested for their validity in classical experiments before they are allowed to be used. One social sustainability rule introduced by Wagner-Pinter and his colleagues states that variables should only be used if their casual relation to the end result can be understood by the person affected. Such a limitation doesn't have to be imposed from the outset; the algorithms in machine learning can also of course be used for data mining at first, as a way of identifying potential reasons for whatever kind of behavior is supposed to be predicted. Ultimately, however, only variables with causal relationships that have been substantiated should find their way into the statistical model. And in the event that data mining does help uncover a surprising new cause for a particular behavior, the insight can be used to improve training for the people making the decisions. A familiarity with the causal chains would also likely render a trained statistical model superfluous; the new knowledge could be stored instead in the form of decision-making rules that were legible to people.

I've promised I would give you some help in deciding when it's necessary to stop and evaluate the broader societal damage a decision-making system might cause. As discussed earlier, it's only necessary to do that for the systems we've identified using the algoscope. If it's also foreseeable that a system will affect a lot of people and that a broad-reaching monopoly exists, it's necessary to analyze the potential for harm from the very beginning. That goes for any state use of ADM systems, but also for any commercial profiling systems and, to a lesser degree, systems for evaluating job applications. It's also a given for any systems used by large online platforms, which have a broad reach and are often distinctly monopolistic in character. It may also be necessary to evaluate systems that affect smaller numbers of people if basic legal rights are at stake. All other systems start out in Class 0; their harm potential should be analyzed the first time doubts arise. This will likely require a central office or bureau where complaints can be taken up. Institutions that are already involved today in accompanying and managing social processes must also be strengthened: unions, consumer protection agencies, public media institutions, courts, and NGOs. In the coming years, they will need their own data scientists and socioinformaticists to gauge software's impact on society—in particular, the impact of artificial intelligence.

Now that we have all our ducks in a row, we need to have a brief talk about Artie, who wants to know what his future holds.

# IT'S TIME FOR "THE TALK" ABOUT STRONG AI

Artie has now accompanied us throughout the entire book as a stand-in for a form of *strong AI*. The term refers to software that equals or even surpasses human ability in nearly every respect, that searches out problems on its own and then examines them systematically for a solution. *Weak AI*, on the other hand, is equipped to handle individual tasks—like the systems described earlier that play chess, recognize images, or are able to convert spoken words into text. To date, then, all we've seen is weak AI. Then there's how my colleague Florian Gallwitz puts it with the decision-making aide illustrated here, "Is it truly AI or not?"

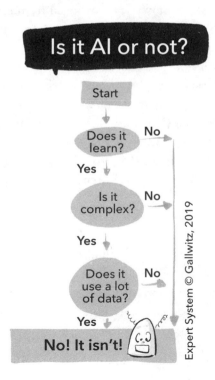

As another colleague, Hannah Bast, once said at an Enquete Commission meeting, "In truth, we've only just transitioned from extremely weak AI to very weak AI." Speaking about artificial intelligence as it appears today, software engineer and theorist Jürgen Geuter has written that "in the end, [it] doesn't exist. Nor is it close to existing. What do exist are efficient statistical systems that have been given a captivating name in order to endow them with a kind of magic. Artificial intelligence is just an advertising slogan."[1] Gallwitz, Bast, and Geuter are all essentially saying the same thing: What we see today has been altogether falsely labeled. It isn't intelligent.

That isn't the same as arguing that strong AI *couldn't* exist, however. And in fact there are many for whom its existence is merely a question of when.

## ARGUMENTS IN FAVOR OF STRONG AI

One such proponent is Jürgen Schmidhuber, a professor and entrepreneur based in Lugano, Switzerland. To work, strong AI needs a master optimization function with an optimization goal by which to measure the success of its actions. Just like the quality metric in machine learning, the optimization goal serves as a compass by which strong AI is able to orient its development. Schmidhuber programs his own robots with an evaluation function he calls *artificial curiosity*. To do so, he uses two learning systems. The first, called the *world model*, attempts to predict things in the world; the second, called *the designer*, attempts to develop sequences of action that improve the robot's optimization function with the help of the world model. The designer is also rewarded when it succeeds in surprising the world model by making a new discovery—something that results in a new pattern to simplify the model currently in use.[2] One example might be introducing a new concept that encapsulates what had to that point been individual cases.

For his part, Schmidhuber is quite certain strong AI will materialize, and that in less than ten years his laboratory will be home to a robot with the intelligence of a capuchin monkey.[3] Nor does it give him great cause for concern as he considers it to be a quasi-natural part of evolution. He works under the assumption that strong AI would quickly propagate itself in outer space to secure the resources necessary for its continued development and doesn't see it as posing any real threat to humanity.

Overall, computer science isn't unanimous about whether strong AI is feasible or not. The most common argument in its favor turns on a sort of atheistic and materialistic stance that is grounded in the natural sciences and championed by many in Silicon Valley.[4] According to this viewpoint, everything that exists on earth is material, with the intellect, soul, motivation, drive, even consciousness in no way separable from it. Such a perspective views these latter phenomena as nothing more than what are called *emergent properties*, which have grown out of the system based on the complexity of the nervous system or structures in the brain, and possibly the rest of the body.[5] No experiment exists to refute the argument; only an attempt at constructing such a consciousness might confirm it. This goal is pursued with what is at times almost religious fervor—not least because it would represent another stepping-stone toward proving that God doesn't exist, similarly to the discoveries of Copernicus, Darwin, or Freud.[6] Many join Antony Garret Lisi in assuming robots will even become our "overlords."[7] The possibility led roboticist Anthony Levandowski to go ahead and found the Way of the Future church as a precautionary measure. As Patrick Beuth writes in an article for *Zeit*, the church "seeks to bring about, accept and worship an AI-based divinity of hardware and software."[8] As always with the internet, it's difficult to tell here what's satire and parody.

Schmidhuber and Levandowski share the opinion that strong AI will come about. Schmidhuber uses two arguments to justify his own research.[9]

First, he doesn't want to decide himself which research holds promise. He cites artificial fertilizers as an example, which he argues has been a driving force in rapid population growth. Today, of course, we know that such fertilizers hold drastic side effects for the planet, a classic example of an invention that is difficult to evaluate in the final instance. Second, he holds development to be an unstoppable force. There are so many people fascinated by the topic today that sooner or later someone will "put the puzzle pieces together. It isn't possible to say: 'We're going to put a stop to it now.'"[10]

In terms of the first question, whether strong AI is even possible to begin with, I have to say the answer is practically irrelevant. The discussion we need to have is whether it *should* exist—and there I hold a very different opinion from my brilliant peers.

Second, it's very likely that inventions exist whose potential risks so far outweigh their benefits that we can all agree they shouldn't be developed—or, if they already have been, that they shouldn't be used.

Third, you can never fully prevent someone from developing something somewhere anyway, and it may be much more difficult to do in this case than with chemical weapons or material that can undergo nuclear fission. But that shouldn't stop us from holding up development of strong AI, because to my mind there are very good reasons for not trying it. The first problem consists of finding an evaluation function with an optimization goal that doesn't have any serious side effects.

## THE TROUBLESOME OPTIMIZATION FUNCTION, OR IS THAT RIGHT?

Remember Fenton, the funny little robotic vacuum that learned to race about backward at full speed? And that it did so only because its optimization function was based on a false assumption that the robot had crash sensors all around it? Such anecdotes aren't at all novel and for the most part are quite entertaining. In one blog post, Christine Barron describes her attempts to teach a bot how to flip pancakes in a computer simulation.[11] On the first try, the optimization function rewarded the amount of time that it took for the pancake to land on the floor.

Proceeding logically, the bot thus learned to toss the pancake as high and far as possible to increase flight time. That's all well and good, but a near (or far!) miss is still a miss. In the simulation, Barron simply adjusted the optimization function as often as it took for the bot to do what was meant intuitively. And as long as everything stays virtual, it's as easy as that.

On the internet, there's a whole list of optimization functions in the area of machine learning that meant well but sadly didn't achieve what they were intended to.[12] It's a lesson computer scientists learn quickly and must take to heart: a computer will only ever do exactly what you tell it to. It can also be a thoroughly satisfying experience, by the way, to learn there's one thing out there at least that will follow my carefully thought out instructions, even if no one else will (here's looking at you, family!). But most of the time, programming is a rather humbling experience, revealing that you've overlooked something yet again.

The examples I've cited up to now have been amusing; crash-prone Roombas and flying pancakes don't exactly stand to cause much harm. Less amusing was a statement released by YouTube employees in April 2019, which asserted that the company's optimization function may have been partially responsible for algorithms especially recommending videos that contained fake news and conspiracy theories.[13] *Bloomberg News* spoke with

more than twenty former and current employees at YouTube who described how they had come up with a host of ideas for measuring the "toxicity" of videos, either to flag them or to not recommend them further. Instead of implementing the ideas, however, and halting the spread of toxic videos, the article states that company management continually referred back to the goal that the company had set for itself, which was reached in 2016: for users to watch at least one billion hours of videos on YouTube daily.[14]

The logic behind the goal is described succinctly by top YouTube employees in John Doerr's book *Measure What Matters*: "In a world where computing power is nearly limitless, 'the true scarce commodity is increasingly human attention' [a slogan coined by Microsoft's Satya Nadella]. When users spend more of their valuable time watching videos on YouTube, they must perforce be happier with those videos. It's a virtuous circle: More satisfied viewership (watch time) begets more advertising, which incentivizes more content creators, which draws more viewership."[15]

This kind of thinking leads to many more people spending much more time with technology than they might wish, a situation that has already been discussed extensively by a number of authors under the term *addictive technology*.[16] What's more, videos that "will open your eyes" to the purported side effects of vaccinations, the conspiracies of the rich and powerful, or the "theory" that Angela Merkel just might be Hitler's daughter keep viewers' eyes glued to the screen for particularly long periods of time.[17] It may well have been YouTube's fixation on its goals for viewership that introduced greater numbers of people to greater numbers of these toxic videos.

As this example and thousands of others show, optimization functions that may be well-intentioned but either haven't been thought through entirely or are one-sided in their formulation can have severe side effects, even on the first couple tries. Yet if we're going to have strong AI, we can't do without a "master optimization function" for the AI to measure its actions.[18] What are the chances that we do manage to develop an optimization function for strong AI that functions as we want it to? This is what Schmidhuber proposes with *artificial curiosity*. In an interview for the *Frankfurter Allgemeine Zeitung* newspaper, Friedmann Bieber and Katharina Laszlo persist in asking Schmidhuber whether there might not be side effects in this case, too—an artificially curious robot pursuing different forms of science or research, for example. Schmidhuber's reply reveals another problem: "Hardly, since we ultimately all inhabit environments with the same physical laws that are worth researching." While he may have a point here, it applies only for a physical world model, and not for the sort of perspectives on cultural and social norms that play a perennial role for human researchers when selecting their next research project. What would stop AI from selecting people as a research object? In general, and especially when it comes to people, our choices in research require social consensus, and this means the decision can't be left to AI as simply as that.

A curiosity function in AI that was centered around people, then, would have to include social goals like preserving *human dignity*, *sustainability*, and *social inclusion*—operationalizing them, in other words, to make them measurable. That, or the AI would have to learn how we as humans would evaluate these goals depending on the situation and incorporate them into our decision-making.

### LOOKY HERE!

Would it be possible to design an optimization function or have machines learn with this broader worldview in mind, one which unequivocally places people front and center so that machines could weigh whatever decisions were necessary for humans as any one of us might? And if that were possible, how would we come up with a way of prioritizing the concepts mentioned previously, in cases where societal goals were in conflict with one another? Say that a decision appears sensible from an ecological standpoint, but a number of individuals must give up their property to implement it: Do you use a questionnaire to figure out the balance of individual aims? And what form of ranking prevails if people disagree?

One online study from the Massachusetts Institute of Technology (MIT) provides a fascinating cautionary tale about what consulting people on societal and ethical questions can bring. In the study, MIT asked people how a self-driving car should behave when it appeared unavoidable that people and/or animals would be hurt. The study included different scenarios, all variations of the trolley problem.[19] In some, the number of people in danger varied, or their age and gender; in one, the car would have to swerve to hit the smaller group of people; in another, it didn't. You can find out how you would approach the different situations yourself on the website: http://moralmachine.net. Figure 43 shows a simplified example from one survey round; for most of us, the decision will likely be a no-brainer.

After evaluating forty million such decisions, the team behind the study, led by author Edmond Awad, published the initial results in the prestigious scientific journal *Nature*.[20] Although the survey isn't representative for a number of reasons, it's still possible to draw some interesting conclusions.[21] A strong preference was shown for one's own species, for example; most of the participants would sooner drive over animals then people. Following that, most tried to protect larger groups and tended to favor shielding younger people over the elderly. Awad and his coauthors noted in this regard that participants thus came to different conclusions, for example, then did the German Ethics Commission on Automated and Connected Driving, led by legal scholar and former federal constitutional judge Udo di Fabio. As Rule 9 of the commission's report states, "Any distinction based on personal features (age, gender, physical or mental constitution) is strictly prohibited. It is also prohibited to offset victims against one another. General programming to reduce the number of personal injuries may be

Figure 43
The self-driving car can neither brake nor swerve in time; it has to strike either the cat or the girl. Which way should it go?

justifiable. Those parties involved in the generation of mobility risks must not sacrifice non-involved parties."[22]

Now for the surprising part: the surveys indicate that preferences vary by geographical location. According to the authors' analysis, countries further to the south show a clear preference for protecting women over men. Like Western countries, protecting the young also matters a great deal to them. Eastern countries, on the other hand, tend to shield the elderly.[23] As I've explained, the survey isn't representative, and people may have participated more than once or simply lied in the study, so we can't take the quantitative results as a basis just like that. But to my mind, the results do provide at least some qualitative evidence for just how varied the ethical preferences are when it comes to this particular topic.

What do we do here? Which kind of ruling principle should guide our optimization function? Should strong AI learn from the people in its surroundings? The example of Microsoft's Tay, the chatbot who was reeducated to become a Nazi (discussed in chapter 8), shows that selecting "teachers" in one's vicinity can lead to disastrously unrepresentative views. Should strong AI try to create a representative survey then? For which region? A country? A continent? What if the AI's decisions have global consequences, but people's preferences differ? Writing in this regard, the authors of the MIT study note that "whereas the ethical preferences of the public should not necessarily be the primary arbiter of ethical policy, the people's willingness to buy autonomous vehicles and tolerate them on the roads will depend on the palatability of the ethical rules that are adopted."[24] How much more will that be the case when it comes to distributing wealth, solving conflicts, or finding just punishments for criminals?

Should strong AI be taught, then, by experts in a field of study that has wrestled with similarly thorny problems for centuries? If so, the question still remains as to how it would be transmitted and who should gain entrance to the illustrious circle of those allowed to set the optimization function or teach the AI. In either scenario, the AI would necessarily only represent the views of a handful of people. As you can see, the question of strong AI also raises questions about governance, politics, and society that are as old as the hills.

Any optimization function would likely first be formulated in the abstract, leaving open the second step of operationalization and all the modeling decisions required—and all the attendant problems described in chapter 1.

Exactly what kind of data will serve as input? Who determines which sensors the AI has and the meanings their signals possess? How exactly will this raw data be converted into a metric for "social participation" or a "humane life"? As the MIT report and many other sociological studies have shown, the necessary steps of selecting data and operationalizing social concepts are both strongly influenced by culture. Even in the event that an optimization function is agreed on in the abstract, only one section of people will feel themselves represented.

The starting prospects for strong AI developing aren't too bright. It requires an optimization function, but if that function is going to be centered on humans, it must be fed a meaningful selection of data and must operationalize social concepts, both of which depend greatly on one's cultural perspective.

Who is AI supposed to learn from in this case? We have to think this through very carefully. For if the optimization is merely well-intended and doesn't encourage the exact form of behavior we consider sensible, then side effects are inevitable. If this is our baseline, only two things could still rescue the notion of a strong AI built around humans. The first is if only minor side effects can be expected from practically every single reasonable-looking optimization function. The second is if we are able to intervene quickly when things threaten to go off the rails.

### THE WELL-MEANING OPTIMIZATION FUNCTION AND THE COLLINGRIDGE DILEMMA

Laws are a written attempt to gain purchase on complex situations by means of punishment and incentive. We also have taxes and tax incentives intended to bring about desired forms of behavior. Let's not forget speed controls with potentially heavy fines for speeding, or emissions trading as a way of lessening the amount of pollutants that are released. Some such efforts have ended up creating what are called *perverse incentives*—that is, incentives exploited by people in unforeseen ways that lead to the exact opposite of what was intended.[25]

In his book *The Logic of Failure: Recognizing and Avoiding Error in Complex Situations*, Bamberg psychologist Dietrich Dörner makes quite clear his conviction that as humans, we aren't capable of gaining an overview of complex systems or controlling them. I share his opinion wholeheartedly:

humans aren't good at controlling complex systems by means of *a single optimization function determined at the outset*. I consider the notion of using a single optimization function itself to be a vain endeavor, in which case our final hope would be to quickly switch out it out in the event of unexpected side effects. This leads me to the Collingridge dilemma.

In a book on technology policy from 1981 called *The Social Control of Technology*, David Collingridge raises the issue that the unintended consequences of applying new technologies often first become recognizable only after they've become so prevalent that they can only be reversed at tremendous economic expense.[26] The classic example here is the internal combustion engine and individualized transportation. So long as there were only a few cars on the road, using benzine and diesel wasn't a problem, and it would have stayed that way for a long time. It is only with the mass spread of cars and the accompanying urban planning, people choosing where they lived in the certainty of being able to drive to work, and the dependency of the GDP on the industry that it has become so difficult to find alternatives. A similar situation is conceivable in the case of strong AI. Digital knowledge can be copied, which means it would likely be difficult to keep every single copy up to date. Moreover, with the current state of technology, the knowledge structure—depending on the size of the problem—would at least in part have to be set back to zero. That is, if the context in which the system is used changes too drastically, the system might need to restart its training from zero. If for example a new law required new information to be taken into consideration when deciding whether to brake a vehicle or not, a neural network would now receive one more input than before. The old one couldn't simply deal with that; the system would be as blank as a newborn baby and would have to be retrained. In this case, too, the Collingridge dilemma would present itself.

To put it another way, I believe the total number of feasible optimization functions for strong AI that are compatible with humans to be infinitesimally small when compared to the total number of feasible optimization functions that would have zero positive effects for the planet and its flora and fauna, including humanity. With such dim prospects for the outcome, I think it would be tantamount to playing Russian roulette to try it. The following diagram gives a visual summary of my own arguments.

Yet maybe the benefits of strong AI far outweigh the risk? Unless you seriously believe that humanity is too brainless to even find the right

## WHY STRONG AI SHOULDN'T EXIST

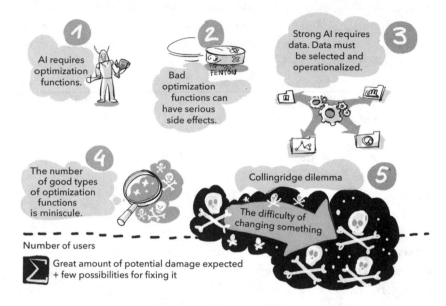

questions, there's no reason to develop strong AI in this form. It's possible to develop individual weak AI systems for any question that we as a society might be curious to pursue. It's also possible to develop weak prototypes that can quickly learn whatever tasks we might set them; we're already seeing that today in software that essentially follows Schmidhuber's method, with two learning systems improving each other.

To my mind, there's no reason to take it one step further.

And with that, gentle reader, I must bid you adieu for now. I've tried my best to give you a sense of why computer scientists find the possibilities of machine learning so thrilling, based on my own journey from natural scientist to data scientist. This enthusiasm holds so long as machine learning isn't used to make decisions about people or their access to basic societal, economic, or ecological resources; in those cases, we have to exercise caution—both to take advantage of the benefits and to avoid the pitfalls.

To give you a part in all this, I've introduced four tools throughout the book: the long chain of responsibility, the OMA principle, the algoscope, and the risk matrix.

The *long chain of responsibility* pointed out all the potential trouble spots when it comes to predicting human behavior. Figure 44 offers one last diagram of it. You can rest assured that most of the questions really are of an ethical or moral nature and that it's both possible and important to involve yourself in the process, be it as a user or as someone affected by what's being decided.

At the same time, the long chain of responsibility demonstrates that even if a system does feature learning components, there's basically nothing to fear from artificial intelligence when it's meant to decide on things without any direct bearing on people or society per se. In most of these situations, the quality metric is self-evident and easily measured; there's no need for a fairness metric because neither people or their participation in society are at stake; and there probably aren't any social concepts that need operationalizing. All this makes the *OMA principle* significantly easier to test for, which requires that the modeling of the question is an accurate fit for the algorithm (including any and all operationalization decisions).

The *algoscope* vindicates itself on the same grounds. At a basic level, artificial intelligence—that is, the algorithms of machine learning—demands

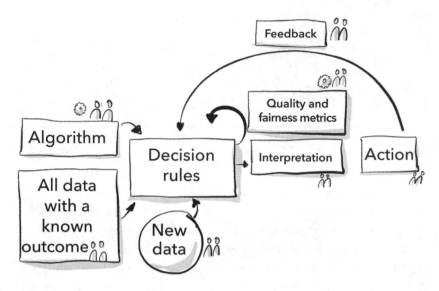

Figure 44
You're wanted backstage: We need your input! The path to better algorithmic
decision-making systems begins with you voicing your opinion and getting involved
anywhere the people icons appear.

our attention either when it *reaches verdicts* (makes decisions about humans)
or *composes verses* (replaces humans in their most essential activities). Not
all these systems need oversight at the *technical level*, however. The algoscope helps you decide whether a particular decision-making system calls
for expanded oversight and regulation. It follows from this that only a relatively small class of software systems need consideration from an ethical
perspective. In the first instance, those are systems that use data to learn
about people's past behavior in order to then draw conclusions about how
others may behave in the future, as well as those that will determine access
to social resources. Here, in other words, it's essentially about algorithmic
systems that reach verdicts.

    In addition to this are systems that are subject to safety regulations, for
example, which also obviously call for some form of technical oversight.
Whether machine learning can even be introduced in such situations is still
being researched. As a way of determining the level of technical oversight
required for decision-making systems that reach verdicts, I introduced you
to a *risk matrix* with five different levels of regulation.

ADM systems that create verses—that is, take over work done by people—will also have serious consequences for education, labor, and social policy that we must meet head on. AI systems that alternatively try to orchestrate our political beliefs, keep us enthralled as users, spy on us, turn us into addicts, or otherwise manipulate us also demand answers in the form of data privacy laws, consumer protection laws, and security policy. All this is obviously notwithstanding the ongoing need to furnish proof of a product's functionality, as might be tested for by an algorithmic safety administration. Ensuring car brakes operate safely comes to mind, or ensuring work safety if a robotic arm is placed next to a human employee on the factory line.

On the strength of several examples, I've shown that an algorithmic safety administration isn't the right place for figuring out when it makes sense to apply ADM systems to complex social processes. That's because the social context itself is what first determines the potential risks, and with them the quality metric. Analyzing the potential degree of harm to the overall *sociotechnical system* requires people who have been educated to do so, which is why at the University of Kaiserslautern we've developed socio-informatics as a course of study that teaches this skill specifically, among others.

More than anything, though, this book was intended to help you see that the system is malleable and that we're capable of acting as citizens. Dividing decision-making systems up into classes, with treatment that varies depending on the need for technical oversight, gives us a clear blueprint for what we can do in this regard and what's needed to make that happen. Below is a summary of typical questions that need answers when an ADM system is in development. All this gives me cause for optimism that in the future we will get a handle on the essential points of the technology.

### A Selection of Ethical Questions to Ask When Developing and Using Algorithmic Decision-Making Systems

*Data:* Which social concepts have been *operationalized*, and how? Overall, what data has been used? How high is the quality? Who's defining the ground truth?

*Methods:* What type of algorithm is being used? Is it a good fit for the quantity of data available, or is it actually too data-hungry? Is it robust enough? Is the resulting statistical model comprehensible to the average person?

*Quality and fairness metrics:* What quality and/or fairness metrics have been used? Who decided on each?

*Data entry:* What possibilities exist for making errors when entering data during operation?

*Interpretation:* How exactly are the results presented? Who's interpreting them? Has the person been trained? Are the values for other quality metrics known? Have the meanings of these values been clearly communicated?

*Action:* Who makes the final decision (the action)? Does the machine make decisions autonomously, or is there a human decision-maker who comes after?

*Feedback:* Is feedback coming from both sides, or not? Is that being measured? How? How does it go into improving the system? There's also the question of the overarching goal: Who set the social goal the machine is supposed to improve on? How is whether that goal has been achieved being measured?

The sort of timescale in which we'll manage to pull that off, on the other hand, is less clear to me. There's always the Collingridge dilemma to consider: not all of a technology's side effects have been discovered so long as it hasn't found wider use. Yet once it's been applied across the board, it becomes difficult to control or get a grip on for that very reason. Facebook, and its use by other countries to influence US political opinion, is a potent example in this regard.

So what can you do? Start right away. Think about how you make your own decisions: What would matter to *you* if your company were to hire a new employee? Talk about it at the workplace: What decisions are you already making? Could data help you to make better decisions? What criteria do you use to decide whether or not you've made a good or a bad decision? Can that be made measurable?

Or think back to the last time you had to make a hard decision in your family, social organization, or circle of friends. What were the values that guided your decision-making together? Can you imagine making those values understandable to a machine?

It's only once you've mulled over what a good decision actually entails, for both yourself and the circles in which you're active, that you'll also be in a position to decide the extent to which machines can and should assist you.

At the end of the day, there's only one way for ethics to find its way into machines, and that runs through you, through me, through us!

## IN CLOSING, A THANK YOU

I'd like to thank all my test readers for their helpful notes. That especially includes my parents, Peter-Hannes Lehmann and Ursula Lehmann-Buss, who have read through my German texts for more than twenty years now, and who as journalists have been instrumental in making my scientific writing understandable. I'd also like to send out a large thanks to Konrad Rauße, Anita Klingel, and Silke Krafft, each of whom carefully and lovingly combed through the details of the text. There's also Forian Eyert and Andrea Hamm at the Weizenbaum Institute for the Networked Society to thank for their detailed comments, as well as ministerial councilor Claudia Bülter, the head of the Secretariat of the Enquete Commission on Artificial Intelligence. On the publishing side of things, I'd like to thank the team at Heyne Verlag for their valuable all-round support, especially Julia Sommerfeld and Laura Sommerfeld.

My thanks to Hannah Leitgeb at the rauchzeichen.agentur for reaching directly for the phone after reading my portrait in the *Süddeutsche Zeitung*. I also owe a great debt of gratitude to the editor of that article, Christina Berndt. Without you, my dear Ms. Berndt, this book would probably still be lying somewhere in a drawer.

I'd also like to thank Tobias Krafft, with whom I've spent the past several years discussing and exploring many of the topics raised here. Along with my husband, in early 2019 Tobias and I founded Trusted AI (https://trusted-ai.com) as a way of acquainting people with the subjects in this book through lectures and workshops. We also accompany state institutions and private companies throughout the process of issuing tenders for, using, and evaluating algorithmic decision-making systems. I'd be delighted to continue exploring these topics together if there's interest!

For the English edition, I would like to thank the publisher. Thank you for all the care that went into this process. Special thanks go to Noah Harley, the translator, who was incredibly creative and thorough in translating all my German puns and anecdotes into English. Thank you. I am also very much touched by the sensitive and in-depth remarks of an anonymous reviewer of the translation. Thanks to both of you for transferring the book's many ideas into English—a feat that none of today's AI translation systems could have done so well.

In a family, anything that comes about does so as a result of teamwork. With that, my greatest thanks go to my husband, who has supported me at every step along my career. The only other reason this book exists is that he made it possible for me to write at night—which meant I couldn't help out with the kids as much as otherwise. Thank you.

## ALGORITHM

An algorithm is a set of instructions for solving a mathematical problem sufficiently detailed and systematic such that with correct implementation (translation into code) by any experienced programmer, a computer will calculate the correct output for any correct input quantity.

## BIG DATA

Big data is a spellbinding term that always refers to very large quantities of data, and most often to batches of data that arise in one context but are evaluated for another. One example is shopping data evaluated for recommendation systems. Other aspects of big data are that it is usually incomplete, may contain errors, and is frequently composed of different sources. This presents a further challenge as it requires understanding which pieces of information are in fact describing the same person. Big data usually has to be processed very quickly as it takes up so much space that it can't be saved over the long term.

## BLOCKCHAIN

I know, I know, the term blockchain didn't show up once in the whole book. But after one lecture, a listener asked me, "I've got the whole AI thing now, but what does any of it have to do with blockchain technology?" So for those of you who may also be wondering, I'll give you a short answer: nothing at all. At the moment, blockchain technology is just another alluring but elusive critter being chased about the global village. Blockchain technology eliminates the middlemen and middlewomen responsible for establishing trust between trading partners in centrally organized institutions—notaries, banks, title registries, and the like. How does it work? Essentially, the entire (global) village is invited to be present for a trade to which you bring a sack clearly marked with a number that is assigned one time and one time only. Once the majority of the village confirms they've seen the exchange, it's recorded into a really really really long document for all to see and copied into the budgets of all the participating villagers. This means the technology consumes a great deal of energy as every single budget has a copy of every single trade made to date. That's how it works at least in the bitcoin universe, but also with most other blockchain variants. New variations looking to lower energy consumption are being proposed constantly. Ultimately what you are doing is exchanging trustworthy intermediaries or brokers for a large number of witnesses. You can thus pursue business without incurring expenses for middlemen (or middlewomen!), who establish trust, but also act on behalf of centralized institutions and can be quite costly.

## CORRELATION

Two properties or patterns of behavior are said to correlate when they are often observed to coincide with one another. Machine learning methods search for correlations when they look for properties that regularly arise alongside whatever behavior or property is supposed to be predicted. Two properties that occur often alongside one another may be linked causally to one another, in the sense of one property causing the other. They might also come from a third property, however, or simply occur simultaneously by coincidence. When something follows causally from another, the two properties always correlate—but the reverse isn't true. While correlation thus follows from causality, causality does not follow from correlation.

## DATA MINING

Data mining refers to the analysis of large quantities of data. The results are then largely used to improve (business) processes. That isn't always a simple matter and is a little bit like mining; you often have to move quite a bit of (data) rubble out of the way in order to come up with a little gold.

## DATA SCIENCE

Data science is the discipline of teaching the methods of data analysis and how to communicate the results. This includes statistics, machine learning methods, and the visualization of data and results. Often, it also includes a working knowledge of how best to build software for supporting people in data analysis (the methods of human-computer interaction).

DIGITALIZATION

Digitalization designates the process of converting information into something a computer is able to process. If the information consists of numbers, it only needs to be stored in a format the computer is able to read. If the information consists of categories or types of relationships, those categories and types must be first assigned numbers and then stored in the computer. If the information concerns (social) concepts like *trust*, *importance*, or *civility*, these must first be made measurable (see *Operationalization*). To do so, a method for measuring has to exist, which is often mathematical in nature but can also occur through a sensor (I include things like video cameras and microphones in this category).

FAIRNESS METRIC

A fairness metric is a mathematical function that evaluates the extent to which different groups within a population are equally affected by decisions. In particular, a fairness metric can measure whether the behavior of people from different groups (e.g., their credit worthiness or risk of recidivism) is predicted correctly at an appropriate rate or whether one group consistently receives a worse rating. The ethical judgment about whether unjustified forms of discrimination are present or not must always be made by people. As with quality metrics, more than two dozen such fairness metrics exist; the social situation to which the algorithmic decision-making system will be applied is what determines which one makes sense. In cases where the decisions either do not immediately affect people or do so indirectly, fairness metrics aren't necessary to assess how good the decision-making system is.

GROUND TRUTH

In machine learning, ground truth is the name given to the "actual" results from the training and test data sets, which the predictions are either based on or compared to, respectively. After first being given the training data set, the algorithm tries to identify properties within the data that appear often alongside the outcome that is to be predicted (e.g., a criminal's recidivism) and that are rarely found outside that outcome (e.g., rarely found among those who don't recidivate). The resulting statistical model is then given the test data set and makes a prediction (i.e., calculates a result based on the decision rules that have been learned), which is then compared to the real result. There are dozens of ways to measure the quality of a prediction against the ground truth (see *Quality metric*), in the same way that dozens of methods exist for judging whether decisions were free of discrimination, or "fair" (see *Fairness metric*).

## HEURISTIC

As opposed to an algorithm, a heuristic is a strategy for finding a viable solution to a problem that also isn't necessarily the best one. Taking the shortest path problem as an example, one heuristic might be this: "Find out which cardinal direction your goal lies in. In each case select a street that you haven't tried out yet and that runs as close as possible in a straight line toward your destination, as the bird flies. If two options exist, choose the more northern of the two. Moving down this street, use this strategy at each intersection you come to. If you wind up in a cul-de-sac or return to a place you've already been, go back to the last intersection where there is a street you still haven't tried out." Such a strategy will find a path if one exists, but it won't necessarily be the shortest.

## IMPLEMENTATION

An algorithm has undergone implementation when it exists in the form of code. Because algorithms are in fact only detailed sets of instructions (see *Algorithm*), they can also exist in other forms, such as text or pseudocode. Pseudocode uses the structural conventions of programming languages and is already quite close to implementation, but in a shortened form that is still easy for people to read.

## MACHINE LEARNING

Machine learning refers to a collection of methods that search among data for patterns that allow for predictions in the future. In most cases, training is accomplished with the aid of a ground truth (see *Ground truth*). This means that a person's data are linked to their behavior in the past—Applicant A was a successful hire, Applicant B was not—and the methods of machine learning then attempt to identify properties that appear frequently with one behavior and infrequently with the other. The patterns discovered in the process are then stored in a statistical model in the form of decision rules. A second algorithm takes these decision rules and is then able to make decisions for new data.

## MODELING

I take the term modeling to mean any simplification or abstraction of a situation that is still precise enough to allow for predictions or analytical conclusions about the situation. Different sorts of modeling were discussed throughout this book:

- Making social concepts measurable, a special form of modeling also referred to as *operationalization*. One example would be one of the many online network services that measure the relevance of a piece of news or a website essentially by how frequently users interact with it.
- Simplifying a situation to end up with a mathematical problem that can be solved with the assistance of a preexisting algorithm. In this case, a model of the world is developed that is simple enough for a classical algorithm to be able to solve.
- Abstracting a model based on different data and their relationships to one another, with reference to an outcome that's intended for analysis with an algorithm that uses machine learning.

## OPERATIONALIZATION

Operationalization is the process of making a (social) concept measurable, and it is always based on a model of the concept. You might measure friendship, for example, by observing how often or for how long two people talk to each other, or whether they're listed as friends on a social media platform. You could also ask both people whether they're friends and measure their friendship by how they reply. Any of these represents one possible operationalization of the term *friendship*. What's more, each comes with advantages and disadvantages. Many are easy to measure but are less precise for it. Others may be more precise but only capture one aspect of friendship.

## QUALITY METRIC

A quality metric is a function that evaluates how successful an (algorithmic) solution to a problem is. Especially when it comes to the algorithms and heuristics of machine learning, it's the standard that determines whether a decision-making system is implemented or not. More than two dozen different quality metrics exist in computer science. Which one is best suited for a task is always determined by the context in which the decision-making system will eventually be applied.

## WEAK AI

Weak AI is software able to replace people in performing a discrete cognitive activity that was previously considered specifically human. Examples include image recognition (What do you see?), playing chess on the computer, or predicting when a building component will need to be replaced.

## STRONG AI

Strong AI refers to software that is equal to or even superior to human ability in nearly every respect.

NOTES

PREFACE

1. The term *nudging* refers to any measures taken to push, or nudge, people's decision-making in a particular direction—using power-saving devices by default, for example. Nudging can make it easier to bring about socially desirable behavior, though in more extreme cases it can be perceived as high-handed. In the business world, it can take the form of manipulating customers.

2. There's even a term for this, a *bias blind spot*.

3. Author D. J. Patil lays claim to the title alongside Jeff Hammerbacher in his wonderfully titled "Data Scientist: The Sexiest Job of the 21st Century." Thomas H. Davenport and D. J. Patil, "Data Scientist: The Sexiest Job of the 21st Century," *Harvard Business Review*, October 2012, 70–76.

4. Anasse Bari, Mohamed Chaouchi, and Tommy Jung, *Predictive Analytics for Dummies* (New York: John Wiley & Sons, 2014), 7–8.

5. Inostix, "How HR Analytics Will Transform the World of Hiring," HRMinfo, October 15, 2014, https://blogs.hrminfo.eu/2014/10/15/how-hr-analytics-will -transform-the-world-of-hiring/.

CHAPTER 1

1. The extent to which the answers were used wasn't at all transparent, by the way. According to one blog post on the website of the company (which today is called equivant), the actual risk assessment was based on only six inputs (https://www .equivant.com/official-response-to-science-advances/). But these inputs seem to have been aggregated from the data described. That at least is the suggestion in the COMPAS handbook, which in paragraph 4.1.2 on page 27 lists "prior criminal history, criminal associates, drug involvement, and early indicators of juvenile delinquency problems" as factors in predicting general recidivism. See https:// www.equivant.com/wp-content/uploads/Practitioners-Guide-to-COMPAS -Core-040419.pdf.

2. Here they were using ROC AUC; what exactly that percentage expresses is explained in chapter 5.

3. Julia Dressel and Hany Farid, "The Accuracy, Fairness, and Limits of Predicting Recidivism," *Science Advances* 4, no. 1 (January 17, 2018), https://advances .sciencemag.org/content/4/1/eaao5580.

4. The German equivalent is a Technical Inspection Association, (a Technischer Überwachungsverein, or TÜV for short).

5. Katharina Zweig, Sarah Fischer, and Konrad Lischka, "Wo Maschinen irren können," *AlgoEthik-Reihe der Bertelsmann-Stiftung* (April 2018), https://doi.org/10 .11586/2018006.

6. Zweig, Fischer, and Lischka, "Wo Maschinen irren können."


CHAPTER 2

1. Don't be too quick to judge: the poor little yeast cells from my thesis shared the same fate as the ones that produced your beer, wine, and last Sunday's rolls!

2. ★Sarcasm off★

3. Sabrina Büttner, Tobias Eisenberg, Eva Herker, Didac Carmona-Gutierrez, Guido Kroemer, and Frank Madeo, "Why Yeast Cells Can Undergo Apoptosis: Death in Times of Peace, Love, and War," *Journal of Cell Biology* 175, no. 4 (November 2006): 521–525.

4. Initially referring to an understanding of the written word, the term *literacy* was later expanded to include understanding problems in mathematics. Today it can mean anything needed to take a critical approach to a given skillset or ability.

5. He wasn't the first, nor will he be the last. https://tylervigen.com/spurious -correlations; see also Tyler Vigen, *Spurious Correlations* (New York: Hachette Books, 2015).

6. See https://tylervigen.com/view_correlation?id=79686. Take your time and play around on the website a little. Whenever you find yourself concocting a story about why this or that correlation might actually be causal in nature, try to come up with a story that works in the opposite direction and think about how you might test it out experimentally.

7. Data from Tyler Vigen's site: http://tylervigen.com/view_correlation?id=31365.

8. It is a hypothesis because the term *computability* can't actually be conclusively defined. What legions of brilliant people have done instead is define various computability models that are intuitively meaningful and shown that they are all capable of computing the exact same things and aren't able to solve other questions. One computability model simulates what a human is capable of calculating with a pen and paper; another simulates what mathematics is able to compute when aided by functions. Because no human has succeeded in defining a computability model that deviates in any way to date, we still hold that people and machines are able to

calculate the same things and that there exists a natural number of questions that can be calculated. Nor will quantum computers do anything to change the fundamental division between "computable" and "noncomputable" functions; they'll just speed up some calculations tremendously.

## CHAPTER 3

1. A couple of years ago I actually received an inquiry for a lecture on the power of logarithms. "Sure, but it will be pretty short," I could imagine myself happily replying. "Their power lies in the base, then grows exponentially!" It would have been a nonstarter, though, an inside joke. Besides, it was obvious that the person had meant to ask about the power of *algorithms*.

2. For those who prefer a more formal expression of the problem, here it is: Sought are the numbers u, v, x, and y such that $u + 2 = v - 2 = x / 2 = y * 2$ and $u + v + x + y = 45$.

3. In a computer, of course, all information is stored in the form of numbers. The difference between numbers and text as forms of input is that there are only different numerals or digits with the former, whereas text tends to be stored more often as symbols. With numbered inputs, the small number of digits that exist can then be used for more efficient methods, if the numbers themselves aren't too long (for those in the know, I'm talking about the radix sort algorithm).

4. See https://www.youtube.com/watch?v=kPRA0W1kECg.

5. I had initially written "a talented two-year-old" here instead, which prompted a furious phone call from my father. "Katharina! I was always at least a talented three-year-old; there's a huge difference. And say hi to my grandson for me." And you know what? He was right, so I changed the text!

6. Yogi Berra, *When You Come to a Fork in the Road, Take It!* (New York: Hyperion Books, 2001), 53. Really though, so far so good—until the day, that is, that my otherwise high-speed train was delayed while picking up other passengers from a stranded train. Once I arrived in Berlin and rushed, months pregnant, out across the main plaza in front of the station, I slipped and fell. Fortunately, my now fashionably torn tights didn't draw any further attention while I was onstage at Parliament!

7. William Edwards Deming also calls it an *operational definition*: "An operational definition is a procedure agreed upon for translation of a concept into measurement of some kind." *The New Economics—For Industry, Government, Education* (Cambridge, MA: MIT Press, 2000), 105.

8. Thomas Misa, "An Interview with Edsger W. Dijkstra," *Communications of the ACM* 53, no. 8 (August 2010): 41–47, https://cacm.acm.org/magazines/2010/8/96632-an-interview-with-edsger-w-dijkstra/fulltext.

9. Here I'm borrowing from Wikipedia's definition of the concept of a revolution. See https://en.wikipedia.org/wiki/Revolution.

10. See https://www1.deutschebahn.com/db-analytics-de-thema-b/content-seite -b3-962226.

11. Misa, "Interview with Edsger W. Dijkstra."

12. Misa, "Interview with Edsger W. Dijkstra."

13. And with that I'd like to salute all of those who make our world a more diverse and colorful place with their own models for relationships, presenting completely new challenges in the theoretical modeling of the problem in the process. Come to think of it, I don't know of an LGBTQ variation on the problem to date . . . maybe something for a bachelor's thesis.

14. Kevin Wack, "'I Lost My Home Because of a Computer Glitch': Wells' Victims Seek Answers," *American Banker*, November 13, 2018, https://www.american banker.com/news/i-lost-my-home-because-of-a-computer-glitch-wells-fargo -victims-seek-answers.

15. Siddarth Cavale, "Wells Fargo Says Internal Error Caused More Home Foreclosures than Expected," Reuters, November 6, 2018. Available online at: https:// www.reuters.com/article/us-wells-fargo-housing/wells-fargo-says-internal -error-caused-more-home-foreclosures-than-expected-idUSKCN1NB23S.

16. Helen Coffey, "Airlines Face Crackdown on Use of 'Exploitative' Algorithm that Splits Up Families on Flights," *Independent*, November 19, 2018, https:// www.independent.co.uk/travel/news-and-advice/airline-flights-pay-extra-sit -together-split-family-algorithm-minister-a8640771.html.

17. The report can be found here: https://publicapps.caa.co.uk/docs/33/CAP 1709_Paidfor_allocated-seating_updateOCT2018.pdf. Of those flying in groups who didn't pay the extra fee, 35 percent were separated from the others in their group. For other airlines, it was 20 percent of passengers or much lower.

18. N. J. Butcher, J. C. Barnett, T. Buckland, and R. M. H. Weeks, *Emergency Evacuation of Commercial Passenger Aeroplanes* (London: Royal Aeronautical Society, April 27, 2018), https://www.aerosociety.com/media/8534/emergency-eva cuation-of-commercial-passenger-aeroplanes-paper.pdf.

19. See Russell Hotten, "Volkswagen: The Scandal Explained," BBC News, December 10, 2015, https://www.bbc.com/news/business-34324772.

20. Donald Knuth and Michael Plass, "Breaking Paragraphs into Lines," *Software— Practice and Experience* 11 (1981): 1119–1184, http://www.eprg.org/G53DOC /pdfs/knuth-plass-breaking.pdf.

21. Knuth and Plass, "Breaking Paragraphs into Lines," 1162.

CHAPTER 4

1. At https://trends.google.com/trends/, you can compare the relative frequency of search queries for up to three terms. To do so, Google Trends sets the highest absolute number of searches for a term as 100 percent, then gives the frequency for the other two as a proportion of that number. Absolute search volumes aren't listed.

2. Visit https://www.internetlivestats.com/google-search-statistics/ for one site where you can see this kind of estimate, though its sources are unclear.

3. Cal Jeffrey, "Taking That Picture of a Black Hole Required Massive Amounts of Data," *Techspot*, April 12, 2019, https://www.techspot.com/news/79637 -taking-picture-black-hole-required-massive-amounts-data.html.

4. Jeffrey, "Picture of a Black Hole."

5. The video is embedded in a tweet: https://twitter.com/NaturePortfolio/status /1116356476161990656. A longer video on YouTube shows many more scientists, all of whom are quite plainly excited: https://www.youtube.com/watch ?v=YNGBIC1zq8c. Both videos were published by the well-regarded science journal *Nature*.

6. Every computer scientist has a slot under their office door so their families can easily slip them XXL pizzas. Coca-Cola and coffee are pumped directly into the office via hose. We can briefly tolerate someone coming into clean once a week while we stretch out on the well-worn sofa to catch a snooze.

7. The challenges were actually quite daunting at the time. It was only with a great deal of finesse, and then only just barely, that I managed to fit the data into my laptop's RAM—that is, the memory from which the computer fetches data to make calculations. The size of a computer's RAM is tremendously important for smooth data processing. It lies very close to the computer's central processing unit (CPU), giving the CPU direct access to the data stored in the memory. This is also where RAM gets its name from—random-access memory, or RAM for short. As soon as any given amount of data becomes so large that it no longer fits, bits and pieces are transferred back and forth between the RAM and the main memory. The transfer between RAM and CPU occurs at high speeds, however; comparatively, the transfer between RAM and the main memory is about as slow as a donkey cart. To give you an order of magnitude, this kind of "swapping" takes about a million times longer than direct access. In 2007, normal RAM was still quite limited, and large amounts of RAM was quite expensive. Even if I had invested the money, though, Java—the programming language I was using—wouldn't reasonably have been able to handle the drive. Which is why it cost me so much effort to cram all the data into the small amount of RAM I had available.

8. This particular quality metric for a prediction is called the root mean square error (RMSE).

9. All the numbers in this paragraph come from the Wikipedia entry for the Netflix Prize, which is worth a read. See https://en.wikipedia.org/wiki/Netflix_Prize.

10. The following numbers all refer to a randomly drawn sample of ten thousand users.

11. That includes "interestingness measures" in the case of transaction rules. See Liqiang Geng and Howard Hamilton, "Interestingness Measures for Data Mining: A Survey," *ACM Computing Survey* 38, no. 3 (2006): 9.

12. Here, see for yourself: https://en.wikipedia.org/wiki/VeggieTales.

13. This joke just barely survived husband quality control—just barely. My husband would like to emphasize that the story is entirely fictional and that he's never been to a guys' movie night.

14. We explain the underlying model in a paper (the *Pretty Woman/Star Wars* result isn't mentioned explicitly; not every single calculation makes it into a scientific paper). Andreas Spitz, Anna Gimmler, Thorsten Stoeck, Katharina Anna Zweig, and Emőke-Ágnes Horvát, "Assessing Low-Intensity Relationships in Complex Networks," *PLOS ONE*, April 20, 2016, https://doi.org/10.1371/journal.pone.0152536.

15. I was also able to furnish mathematical proof that the first model I discussed for evaluating product data will always yield false results when a large variance predominates for both the products and the evaluations—in other words, if products enjoy very different levels of popularity and customers vary greatly in their rating behavior. This is often the case, however, which means that the first model shouldn't have been as popular as it was for so many years purely for theoretical considerations. Katharina Anna Zweig and Michael Kaufmann, "A Systematic Approach to the One-Mode Projection of Bipartite Graphs," *Social Network Analysis and Mining* 1 (2011): 187–218.

16. See also Katharina Anna Zweig, "Good versus Optimal: Why Network Analytic Methods Need More Systematic Evaluation," *Central European Journal of Computer Science* 1 (2011): 137–153.

17. Meanwhile, how many stars you award a film is altogether beside the point as far as Netflix is concerned. What really counts is how long you watched! When did you exit the proverbial theater? How many episodes of a series did you watch in a row at what time? As you can see, these days the data is even bigger and more finely grained.

18. Or the mischievous dolphin-shaped vibrator in Marc-Uwe Kling's *Qualityland* (New York: Grand Central Publishing, 2020), which I highly recommend.

19. As of February 28, 2019, it came up as recommendation 75 on the "Customers who bought this item also bought" list. See https://www.amazon.de/Theo retische-Information-gefasst-Uwe-Sch%C3%B6ning/dp/3827418240.

20. Recommendation 70 on the same "Customers who bought this item also bought" list.

21. Toby Walsh, *Machines That Think: The Future of Artificial Intelligence* (Amherst, NY: Prometheus Books, 2018), 42.

22. Janet Burns, "Tinder Has Been Raided for Research Again, This Time to Help AI 'Genderize' Faces," *Forbes*, May 2, 2017, https://www.forbes.com/sites /janetwburns/2017/05/02/tinder-profiles-have-been-looted-again-this-time-for -teaching-ai-to-genderize-faces.

23. Robert Hackett, "Researchers Caused an Uproar by Publishing Data from 70,000 OkCupid Users," *Fortune*, May 18, 2016, https://fortune.com/2016/05/18 /okcupid-data-research.

24. Kate O'Neill, "Facebook's '10 Year Challenge' Is Just a Harmless Meme— Right?," *WIRED*, January 15, 2019, https://www.wired.com/story/facebook -10-year-meme-challenge.

25. Bill Hart-Davidson makes the same point in a Facebook post: https://m.face book.com/story.php?story_fbid=10113999199234334&id=2364532.

26. The last living northern white rhino in Kenya had an account on Tinder, for example. Something like that can really rattle an algorithm. John Bacon, "Swipe Right! Last Male Northern White Rhino Joins Tinder," CNBC, April 26, 2017, https://www.cnbc.com/2017/04/26/swipe-right-last-male-northern-white -rhino-joins-tinder.html.

27. Who knows—maybe that mixer in your kitchen is spying on you. Security experts in France discovered that Lidl's Thermomix-Kilon contained a microphone. The microphone came deactivated and was only one part of the built-in tablet, but the tablet itself was full of security holes that made it hackable. Nice work. It's for reasons like this that devices connected to the internet simply aren't worth recommending. See "Smart Mixer from Lidl France Has 'Secret' Microphone," *Connexion*, June 14, 2019, https://www.connexionfrance.com /French-news/Smart-Monsieur-Cuisine-Connect-mixer-from-Lidl-France-has -secret-microphone.

28. The following study, for example: Archana Vijaya, Shyma Kareem, and Jubilant Kizhakkethottam, "Face Recognition across Gender Transformation Using SVM Classifier," *Procedia Technology* 24 (2016): 1366–1373.

29. James Vincent, "Transgender YouTubers Had Their Videos Grabbed to Train Facial Recognition Software," *The Verge*, August 22, 2017, https://www.theverge .com/2017/8/22/16180080/transgender-youtubers-ai-facial-recognition-dataset.

30. If you haven't seen it yet, here's the URL: https://www.youtube.com/watch?v=cQ54GDm1eL0.

31. Thomas Davenport and D. J. Patil, "Data Scientist: The Sexiest Job of the 21st Century," *Harvard Business Review*, October 2012, https://hbr.org/2012/10/data-scientist-the-sexiest-job-of-the-21st-century.

32. Davenport and Patil, "Data Scientist."

33. Spitz et al., "Assessing Low-Intensity Relationships."

34. Stefan Uhlmann, Heiko Mannsperger, Jitao David Zhang, Emőke-Ágnes Horvat, Christian Schmidt, Moritz Küblbeck, Frauke Henjes, et al., "Global MicroRNA Level Regulation of EGFR-Driven Cell-Cycle Protein Network in Breast Cancer," *Molecular Systems Biology* 8 (2012): 570, https://doi.org/10.1038/msb.2011.100.

## CHAPTER 5

1. Florian Gallwitz, "Auch 2029 wird es keine Künstliche Intelligenz geben, die diesen Namen verdient," *GQ*, December 14, 2018, https://www.gq-magazin.de/auto-technik/article/auch-2029-wird-es-keine-kuenstliche-intelligenz-geben-die-diesen-namen-verdient.

2. For other examples of translation gone wrong, see https://www.lingualinx.com/blog/the-funniest-examples-of-translation-gone-wrong.

3. In this case, I'm speaking only about AI systems that use a supervised learning algorithm as their learning component, where experiences are assigned clear categories or rankings that the systems are then meant to learn.

4. No kitty-cats came to any harm in the making of this example, and the names of the innocent have been protected.

5. Stephen Milborrow, "Tree Titanic Survivors," Wikimedia Commons, last revised September 10, 2020, https://commons.wikimedia.org/w/index.php?curid=14143467.

6. The Kaggle website at https://www.kaggle.com serves as a platform for data sets, companies, and data scientists. The *Titanic* dataset can be found at https://www.kaggle.com/c/titanic. The page bears the somewhat cynical title "Titanic—Machine Learning from Disaster. Start Here! Predict Survival on the Titanic and Get Familiar with ML Basics."

7. Pedro Domingos, "A Few Useful Things to Know about Machine Learning," *Communications of the ACM* 55, no. 10 (2012): 78–87.

8. It's ridiculous, I know. I lived in a co-op during my days as a student at Tübingen—everything was organic, and every expense calculated down to the penny. Just don't ask about the five-page agreement that my graduate students

drew up when we all chipped in to buy the Italian espresso machine. Compared to that, refinancing the washing machine was a breeze!

9. If you're interested, take a look at a short teaser we made for a project: https://www.youtube.com/watch?v=z_sD9Dj35J0. (That character at the end of the URL is a number, not a letter!)

10. I'm presenting the method in a slightly simplified manner. The real outcome of a support vector machine is a dividing line that needs to satisfy slightly more complicated properties.

11. Behind every "data point," of course, is a human being—an applicant in this case. But when speaking at this level, that fact can often go missing.

12. Cassie Kozyrkov, "The First Step in AI Might Surprise You," Hacker Noon, October 12, 2018, https://hackernoon.com/the-first-step-in-ai-might-surprise-you-cbd17a35708a.

13. Custard Smingliegh (@smingleigh), "I hooked a neural network up to my Roomba. I wanted it to learn to navigate without bumping into things, so I set up a reward scheme to encourage speed and discourage hitting the bumper sensors," Twitter, November 7, 2018, 4:18 p.m., https://twitter.com/smingleigh/status/1060325665671692288.

14. William Blackstone, *Commentaries on the Laws of England*, vol. 4 (London: Sweet, Maxwell & Son, 1844), 358.

15. Dick Cheney, interview with Chuck Todd, *Meet the Press*, NBC, December 14, 2014, https://www.nbcnews.com/meet-the-press/meet-press-transcript-december-14-2014-n268181.

16. The idea of juxtaposing these two quotes comes from a lecture by Cris Moore that is definitely worth a watch, about the limits of computers in science and society. It's online at https://www.youtube.com/watch?v=Sg2jtEY6qms.

17. There are a whole host of studies about the software, with values that always hover around 70 percent—depending on the group being examined. One of those studies was published by the company then called Northpointe (now equivant) based on a set of public data that anyone can follow. William Dieterich, Christina Mendoza, and Tim Brennan, *COMPAS Risk Scales: Demonstrating Accuracy Equity and Predictive Parity* (Traverse City, MI: Northpointe Inc. Research Department, July 8, 2016), http://go.volarisgroup.com/rs/430-MBX-989/images/ProPublica_Commentary_Final_070616.pdf.

18. That follows directly from the definition—you can read more at the Wikipedia page, for example: https://en.wikipedia.org/wiki/Receiver_operating_characteristic.

19. Dietrich, Mendoza, and Brennan, *COMPAS Risk Scales*. Tables A1 through A4 in the appendix designate ten different threshold values for the positive predictive value under the label PV+.

20. That's right, SKYNET, just like in the *Terminator* films.

21. The slides are available at https://theintercept.com/document/2015/05/08 /skynet-courier.

22. Ahmad Zaidan, "I Am a Journalist, Not a Terrorist," *Al Jazeera*, May 15, 2015, https://www.aljazeera.com/opinions/2015/5/15/al-jazeeras-a-zaidan-i-am-a -journalist-not-terrorist.

23. For a look at the current data, see Angela Helm, "While Stop & Frisk Has Decreased Significantly in NYC, Young Men of Color Are Still Hit Hardest: Report," *The Root*, March 14, 2019, https://www.theroot.com/while-stop-frisk -has-decreased-substantially-in-nyc-1833294359.

24. The text reads as follows: "States should incorporate the application of risk assessment instruments to individuals throughout the criminal justice process— including in the pre-trial process, sentencing process, and parole and probation decisions." Quoted in *Smart Reform Is Possible: States Reducing Incarceration Rates and Costs While Protecting Communities* (New York: American Civil Liberties Union, August 2011), https://www.aclu.org/files/assets/smartreformispossible.pdf.

25. For those of you who have already forgotten, here's the link to a glorious video by Scooter for his 1995 hit single, "Faster, Harder, Scooter": https://www .youtube.com/watch?v=j0LD2GnxmKU.


CHAPTER 6

1. Here, for example: https://www.nps.gov/subjects/aknatureandscience/wildlife marineseals.htm.

2. The challenge website can be found at https://image-net.org/challenges/LSV RC/index.php.

3. See https://farm1.static.flickr.com/10/13160739_05cd2aeed5.jpg.

4. The image shows a captcha-like puzzle. The source for the original by Wikimedia user Martin is https://commons.wikimedia.org/w/index.php?curid=18112609.

5. See https://www.ted.com/talks/luis_von_ahn_massive_scale_online_collabor ation?language=en.

6. Nor is it strictly necessary to do this any longer; a risk value can also simply be sent back to the website operators, who then decide what to do themselves. There's a video about reCAPTCHA version 3.0 from Google at https://www.youtube .com/watch?v=tbvxFW4UJdU&t=145s.

7. To avoid data manipulation by those lazy enough to just click anywhere on a reCAPTCHA puzzle, the same puzzle is given to multiple people within a short window, against which people's results are then compared. If they don't agree, they get a new puzzle.

8. See https://developers.google.com/recaptcha.

9. See https://youtube.com/watch?v=fsF7enQY8uI.

10. H. A. Haenssle, C. Fink, R. Schneiderbauer, F. Toberer, T. Buhl, A. Blum, A. Kalloo, et al., "Man against Machine: Diagnostic Performance of a Deep Learning Convolutional Neural Network for Dermoscopic Melanoma Recognition in Comparison to 58 Dermatologists," *Annals of Oncology* 29 (2018): 1836–1842.

11. Tom Simonite, "When It Comes to Gorillas, Google Photos Remains Blind," *WIRED*, January 11, 2018, https://www.wired.com/story/when-it-comes-to-gorillas-google-photos-remains-blind.

12. The TEDx Talk series brings together exceptional figures from the worlds of "technology, entertainment and design" to speak at a high-octane conference. The talk cited here is Joy Buolamwini, "How I'm Fighting Bias in Algorithms," TEDx Talk from TEDxBeaconStreet, November 2016, https://www.ted.com/talks/joy_buolamwini_how_i_m_fighting_bias_in_algorithms/.

13. For example, see Adam Smith, "'Racist' Soap Dispensers Don't Work for Black People," *Metro*, July 13, 2017, https://metro.co.uk/2017/07/13/racist-soap-dispensers-dont-work-for-black-people-6775909/; and Max Plenke, "The Reason This 'Racist Soap Dispenser' Doesn't Work on Black Skin," *Mic*, September 9, 2015, https://mic.com/articles/124899/the-reason-this-racist-soap-dispenser-doesn-t-work-on-black-skin#.bveNTn5Qf.

14. See, for example, Caroline Criado Perez, *Invisible Women: Exposing Data Bias in a World Designed for Men* (London: Chatto & Windus, 2019).

15. The quote comes from an article by Stephanie Dutchen, "The Importance of Nuance," *Harvard Magazine*, Winter 2019, https://hms.harvard.edu/magazine/artificial-intelligence/importance-nuance.

16. Chris Anderson, "The End of Theory: The Data Deluge Makes the Scientific Method Obsolete," *WIRED*, June 23, 2008, https://www.wired.com/2008/06/pb-theory.

17. Nassim Nicholas Taleb, *The Black Swan: The Impact of the Highly Improbable* (New York: Random House, 2012).

18. Cathy O'Neil, *Weapons of Math Destruction* (New York: Crown, 2016). See also Safiya U. Noble, *Algorithms of Oppression: How Search Engines Reinforce Racism* (New York: United Press, 2018); Eli Pariser, *The Filter Bubble: What the Internet Is Hiding from You* (London: Penguin Books, 2012); Yvonne Hofstetter, *Sie wissen alles* (Munich: Penguin Books, 2016); or Shoshana Zuboff, *The Age of Surveillance*

*Capitalism: The Fight for a Human Future at the Frontier of Power* (New York: Public Affairs, 2019).

## CHAPTER 8

1.  On side effects, among others, see Safiya U. Noble, *Algorithms of Oppression: How Search Engines Reinforce Racism* (New York: United Press: 2018); Virginia Eubanks, *Automating Inequality: How High-Tech Tools Profile, Police and Punish the Poor* (London: St. Martin's Press, 2018); Sara Wachter-Boettcher, *Technically Wrong: Sexist Apps, Biased Algorithms, and Other Threats of Toxic Tech* (New York: W. W. Norton, 2017). On third-party manipulation, one good example, though by no means the only, is Macedonian youth's interference with the 2016 US presidential election. In a country grappling with high rates of youth unemployment, youngsters noticed that copying scandalous articles written against Hillary Clinton drew so much web traffic to their sites they could turn it to their profit. For more, see Samanth Subramanian, "Inside the Macedonian Fake-News Complex," *WIRED*, February 15, 2017, https://www.wired.com/2017/02/veles-macedonia-fake-news. See also Dan Tynan, "How Facebook Powers Money Machines for Obscure Political 'News' Sites," *The Guardian*, August 24, 2016, https://www.theguardian.com /technology/2016/aug/24/facebook-clickbait-political-news-sites-us-election -trump. On the legal challenges to data protection, see Wolfie Christl and Sarah Spiekermann, *Networks of Control: A Report on Corporate Surveillance, Digital Tracking, Big Data & Privacy* (Vienna: Facultas Verlags- und Buchhandels AG, 2016).

2.  See https://www.ki-strategie-deutschland.de/files/downloads/Fortschreibung _KI-Strategie_engl.pdf.

3.  Here's an example (German language only): https://www.youtube.com/watch ?v=Q1Kh91kLafU.

4.  Birgit Hippeler and Heike Korzillus, "Arztberuf: Die Medizin wird weiblich," *Deutsches Ärzteblatt* 105, no. 12 (2008): 609–612.

5.  The statistic is taken from the imprisonment rate given in the World Prison Brief, hosted by the Institute for Criminal Policy Research at Birbeck College, University of London and available online at https://www.prisonstudies.org/world -prison-brief-data.

6.  E. Ann Carson, *Prisoners in 2016* (Washington, DC: Bureau of Justice Statistics, January 2018), https://bjs.ojp.gov/library/publications/prisoners-2016. Table 6 shows the proportions have held relatively steady since 2006.

7.  *Smart Reform Is Possible: States Reducing Incarceration Rates and Costs While Protecting Communities* (New York: American Civil Liberties Union, August 2011), https://www.aclu.org/files/assets/smartreformispossible.pdf.

8. The court opinion makes for a fascinating read (German only): http://www
.justiz.nrw.de/nrwe/ovgs/vg_gelsenkirchen/j2016/1_K_3788_14_Urteil
_20160314.html.

9. Jeffrey Dastin, "Amazon Scraps Secret AI Recruiting Tool That Showed Bias
against Women," Reuters, October 10, 2018, https://www.reuters.com/article
/us-amazon-com-jobs-automation-insight/amazon-scraps-secret-ai-recruiting
-tool-that-showed-bias-against-women-idUSKCN1MK08G.

10. Dastin, "Amazon Scraps Recruiting Tool." The reasons for this are many and
complex. Significantly fewer women choose careers in technology, and those that
do often quickly head for the exits. But my point in this book isn't to explore the
reasons for unequal staff makeup; it's to look at how algorithmic decision-making
systems should be trained. How we answer this question will in turn hold implications for whether inequality is strengthened, preserved, or balanced out. What I'm
calling for is that we look at the question in terms of company and/or social policy,
rather than as one to be left to small teams of developers.

11. Dastin, "Amazon Scraps Recruiting Tool."

12. Sonia Paul, "Voice Is the Next Big Platform, Unless You Have an Accent,"
*WIRED*, March 20, 2017, https://www.wired.com/2017/03/voice-is-the-next
-big-platform-unless-you-have-an-accent.

13. Rachel Tatman, "Google's Speech Recognition Has a Gender Bias," *Making
Noise & Hearing Things* (blog), July 12, 2016, https://makingnoiseandhearingthings
.com/2016/07/12/googles-speech-recognition-has-a-gender-bias. Her post cites
other (older) studies with similar findings.

14. Rachel Tatman, "How Well Do Google and Microsoft and Recognize Speech
across Dialect, Gender and Race?," *Making Noise & Hearing Things* (blog), August
29, 2017, https://makingnoiseandhearingthings.com/2017/08/29/how-well-do
-google-and-microsoft-and-recognize-speech-across-dialect-gender-and-race/.

15. See https://www.youtube.com/watch?v=NMS2VnDveP8.

16. "Irish-Born Native English Speaker Left in Visa Limbo after Low Score in
Voice Recognition Test," Australian Associated Press, August 8, 2017, https://
www.abc.net.au/news/2017-08-09/voice-recognition-computer-native-english
-speaker-visa-limbo/8789076.

17. Caroline Criado Perez, *Invisible Women: Exposing Data Bias in a World Designed
for Men* (London: Chatto & Windus, 2019).

18. World Economic Forum Global Future Council on Human Rights 2016–
2018, *How to Prevent Discriminatory Outcomes in Machine Learning* (Cologny, Switzerland: World Economic Forum, March 2018), https://www.weforum.org
/whitepapers/how-to-prevent-discriminatory-outcomes-in-machine-learning.

19. A tweet is a short message posted on the Twitter platform. Users can "like" any tweet whose message they want to support by clicking on a small heart icon. They can also pass the tweet's message on to their own circle of followers by "retweeting."

20. James Vincent, "Twitter Taught Microsoft's AI Chatbot to Be a Racist Asshole in Less than a Day," *The Verge*, March 24, 2016, https://www.theverge .com/2016/3/24/11297050/tay-microsoft-chatbot-racist.

21. You can view the trailer at https://www.youtube.com/watch?v=iGCGhD8i -o4. Director Moritz Riesewieck wrote a book about it (German language only): *Digitale Drecksarbeit* (Munich: dtv, 2017).

22. Of course, a developer team might also purposefully program racist, sexist, or in other ways discriminatory decision-making rules directly into the system. I know of no such cases in the realm of machine learning to date, nor does the UNICEF report (Kochi 2017) consider it likely. Still, it's conceivable.

23. This is how it's explained in Google's FAQ. See https://support.google.com /accounts/answer/27442?visit_id=636897172188013978-778214861&p=gender &hl=en&rd=1#1#gender&zippy=%2Cgender.

24. "We also found that setting the gender to female resulted in getting fewer instances of an ad related to high paying jobs than setting it to male." Amit Datta, Michael Tschantz, and Anupam Datta, *Automated Experiments on Ad Privacy Settings: A Tale of Opacity, Choice, and Discrimination* (March 18, 2015), https://arxiv.org /pdf/1408.6491.pdf. The phrasing here deserves criticism, by the way; if you read the article, you'll find that the advertisement in question offered a training session that would (presumably) result in a high-paying job. Yet the authors' finding has often been cited as though the study showed that men see better job listings, which wasn't the case.

25. In the study, gender was modeled in binary fashion; the possibility that queer or transgender people might receive different ads wasn't examined.

26. The press release was published at https://civilrights.org/2018/07/30/more -than-100-civil-rights-digital-justice-and-community-based-organizations-raise -concerns-about-pretrial-risk-assessment.

27. Julia Angwin, Jeff Larson, Surya Mattu, and Lauren Kirchner, "Machine Bias," ProPublica, May 23, 2016, https://www.propublica.org/article/machine -bias-risk-assessments-in-criminal-sentencing.

28. Northpointe is now named equivant.

29. Jon Kleinberg, Sendhil Mullainathan, and Manish Raghavan, "Inherent Trade-Offs in the Fair Determination of Risk Scores," in *8th Innovations in Theoretical Computer Science Conference: ITCS 2017, January 9–11, 2017—Berkeley, CA, USA* (Saarbrücken/Wadern, Germany: Dagstuhl, 2017), 43:1–23.

30. See also Katharina Zweig and Tobias Krafft, "Fairness und Qualität algorithmischer Entscheidungen," in *(Un)berechenbar? Algorithmen und Automatisierung in Staat und Gesellschaft*, ed. Resa Mohabbat Kar, Basanta Thapa, and Peter Parycek (Berlin: Kompenzzentrum Öffentliche IT, 2018), 204–227.

31. World Economic Forum Global Future Council on Human Rights 2016–2018, *How to Prevent Discriminatory Outcomes in Machine Learning* (Cologny, Switzerland: World Economic Forum, March 2018), https://www.weforum.org/whitepapers/how-to-prevent-discriminatory-outcomes-in-machine-learning.

32. Gerd Gigerenzer, *Calculated Risks: How to Know When Numbers Deceive You* (New York: Simon & Schuster, 2002).

## CHAPTER 9

1. *Filter bubble* refers to the torrent of news pieces algorithms expose us to that confirm opinions we already hold. Eli Pariser worked out the concept in his book *The Filter Bubble: What the Internet Is Hiding from You* (New York: Penguin Press, 2011). It's a difficult concept to verify, in particular because the data is missing. *Echo chambers* refers to groups of friends and acquaintances that share a person's opinion and thus echo the person's views. Echo chambers can also be brought into being and reinforced by algorithms and are just as difficult to research as a concept.

2. Tobias Krafft and Katharina Zweig, *Transparenz und Nachvollziehbarkeit algorithmischer Entscheidungssysteme—ein Regulierungsvorschlag* (Federation of German Consumer Organizations, 2019), German language only, https://www.vzbv.de/site/default/files/downloads/2019/05/02/19-01-22_zweig_krafft_transparenz_adm-neu.pdf.

3. Katharina Zweig, Sarah Fischer, and Konrad Lischka, "Wo Maschinen irren können," *AlgoEthik-Reihe der Bertelsmann-Stiftung* (April 2018), https://doi.org/10.11586/2018006.

## CHAPTER 10

1. "The Use of Pretrial 'Risk Assessment' Instruments: A Shared Statement of Civil Rights Concerns," press release from more than 110 groups, including the ACLU, 2018, http://civilrightsdocs.info/pdf/criminal-justice/Pretrial-Risk-Assessment-Full.pdf.

2. You can read about all of this in a detailed report (German language only) from the company Synthesis Forschung GmBH, which was responsible for developing the underlying ADM system. Jürgen Holl, Günter Kernbeiß, and Michael Wagner-Pinter, *"Das AMS-Arbeitsmarketchancen-Modell,"* *Konzeptunterlage zur AMS-Software* (Vienna: AMS Österreich, October 2018), https://www

.ams-forschungsnetzwerk.at/downloadpub/arbeitsmarktchancen_methode_%20 dokumentation.pdf.

3. Holl, Kernbeiß, and Wagner-Pinter, *Das AMS-Arbeitsmarktchancen-Modell*. Logistic regression is one of the simplest methods out there for learning decision rules based on data, by the way; any number of computer scientists might question whether it even belongs to the field of AI. But when used for predicting human behavior, at least, it does count, as all the problems I'm discussing pertain even with such a straightforward method.

4. Österreichischer Rundfunk (the Austrian Public Broadcast company), "Die Grenzen des AMS-Algorithmus," January 18, 2019, German language only, https://orf.at/stories/3108185.

5. Holl, Kernbeiß, and Wagner-Pinter, *Das AMS-Arbeitsmarktchancen-Modell*.

6. Questions like whether the information is binary (gender) or a whole number (e.g., age) or categorial (e.g., various educational degrees).

7. Wagner-Pinter and his team actually had access to even more properties. If they had used them all, however, the data set would have been split up into so many groups that each individual group would have held too few people. The calculation is the same as before: one property with two possibilities makes two groups, two properties with two possibilities each makes four groups, three properties makes eight, and so on. Yet the number of data points stays the same throughout, which means that a large number of properties would have made each group so small as to make any kind of statistical analysis impossible. Statistics always needs enough people in a given group in order to arrive at any kind of statement. This explains why it was necessary to limit it to twenty-two properties only. Which properties made the cut was decided ahead of time using data analysis.

8. Jürgen Holl, Günter Kernbeiß, and Michael Wagner-Pinter, *Personenbezogene Wahrscheinlichkeitsaussagen ('Algorithmen')—Stichworte zur Socialverträglichkeit* (Vienna: Synthesis Forschung, May 9, 2019), http://www.synthesis.co.at/images /Personenbezogene_Wahrscheinlichkeitsaussagen_Algorithmen_Mai2019.pdf.

CHAPTER 11

1. Jürgen Geuter, "Nein, Ethik kann man nicht programmieren," *Zeit*, November 27, 2018, https://www.zeit.de/digital/internet/2018-11/digitalisierung-mythen -kuenstliche-intelligenz-ethik-juergen-geuter.

2. Schmidhuber has written a large number of technical articles about this, or you can watch an edited clip of one of his lectures at https://www.youtube.com /watch?v=Ipomu0MLFaI.

3. Friedmann Bieber and Katharina Laszlo in conversation with Dr. Jürgen Schmidhuber, "Intelligente Roboter werden vom Leben fasziniert sein," *Frankfurter Allgemeine Zeitung*, December 1, 2015, https://www.faz.net/-i9n-8ata1.

4. I invite anyone who considers this statement to be an overexaggeration to consult John Brockman, ed., *What to Think about Machines That Think* (New York: Harper Perennial, 2015). There you'll find chapters such as Martin Rees, "Organic Intelligence Has No Long-Term Future"; Frank Tipler, "If You Can't Beat 'Em, Join 'Em"; and Antony Garrett Lisi, "I, for One, Welcome Our Machine Overlords."

5. *Emergence* is the notion that interactions between the objects of a system translate into measurable properties at each higher level of remove. Cars can interact with each other on a highway, for example, in such a way that at the model's subsequent level—transportation as a whole—traffic jams can emerge for no apparent reason. At the level of the individual car, on the other hand—itself a system of parts that act on each other to produce the emergent property of driving—the phenomenon of traffic can't be comprehended. For more, see the lovely little explanatory video on the mysterious phantom traffic jam at https://www.youtube.com/watch?v=goVjVVaLe10.

6. Freud himself designated these discoveries *narcissistic wounds*—the cosmological wound, the biological wound, and, in the case of his own discoveries, the psychological wound that we do not stand at the center of things and are only the temporary result of an ongoing process of evolution, with all of the specious behavior that entails. Today I would add a fourth type inflicted by behavioral scientists: the rationalist-economic wound, according to which our decisions are often irrational.

7. Lisi, "I, For One, Welcome Our Machine Overlords," in *What to Think about Machines That Think*, 22–24.

8. Patrick Beuth, "Man kann Kirche nicht ohne KI schreiben," *Zeit*, November 18, 2017, https://www.zeit.de/digital/internet/2017-11/way-of-the-future-erste-kirche-kuenstliche-intelligenz/.

9. From Schmidhuber, "Intelligente Roboter werden vom Leben fasziniert sein."

10. Schmidhuber, "Intelligente Roboter werden vom Leben fasziniert sein."

11. There's a project description at https://web.archive.org/web/201808311953 25/https://connect.unity.com/p/pancake-bot.

12. See https://docs.google.com/spreadsheets/u/1/d/e/2PACX-1vRPiprOaC3 HsCf5Tuum8bRfzYUiKLRqJmbOoC-32JorNdfyTiRRsR7Ea5eWtvsWzuxo8 bjOxCG84dAg/pubhtml.

13. Mark Bergen, "YouTube Executives Ignored Warnings, Letting Toxic Videos Run Rampant," *Bloomberg News*, April 2, 2019, https://www.bloomberg.com/news/features/2019-04-02/youtube-executives-ignored-warnings-letting-toxic-videos-run-rampant.

14. John Doerr, *Measure What Matters: OKRs, the Simple Idea That Drives 10x Growth* (New York: Portfolio, 2018).

15. From Doerr, *Measure What Matters*, 161.

16. See, for example, Adam Alter, *Irresistible: The Rise of Addictive Technology and the Business of Keeping Us Hooked* (New York: Penguin Press, 2018); or Tim Wu, *The Attention Merchants: The Epic Struggle to Get Inside Our Heads* (New York: Vintage Books, 2017).

17. For more on this, read Christian Alt and Christian Schiffer's truly enjoyable and otherwise highly terrifying book about conspiracy theories: *Angela Merkel ist Hitlers Tochter* (Munich: Carl Hanser Verlag, 2018); German language only. It was last year's favorite gift to colleagues.

18. As humans, our "optimization function" is hardwired into our neurons, embedded in our bodies, and regulated by the release of hormones or nerve signals. It's dubious whether this sort of complex functioning, which is also individual and dynamic, could truly be learned.

19. The Wikipedia page on the trolley problem provides an excellent summary of the state of things: https://en.wikipedia.org/wiki/Trolley_problem.

20. Edmond Awad, Sohan Dsouza, Richard Kim, Jonathan Schulz, Joseph Henrich, Azim Shariff, Jean-François Bonnefon, and Iyad Rahwan, "The Moral Machine Experiment," *Nature* 563 (2018): 59–64, https://www.americaninno .com/wp-content/uploads/2017/05/The-MM-Experiment.pdf.

21. For one thing, this has to do with the fact that participants decided to participate themselves (self-selection). On the other hand, it can't be ruled out that some people didn't participate multiple times. It becomes clear how skewed the data set is if you look at the demographic distribution of study participants: many hold academic degrees, 70 percent are men, and the annual income is low. This information leads one to the conclusion that many of the participants must have been students. The internet is a scarce resource in many countries, which leads to a further distortion of the opinions gathered. For more, see the resources at the Supplementary Information link provided at the bottom of the article's site: https://doi .org/10.1038/s41586-018-0637-6. The results are nonetheless still highly intriguing, as they point to a high degree of cultural diversity.

22. *Ethics Commission: Automated and Connected Driving* (German Federal Ministry of Transport and Digital Infrastructure, June 2017), https://www.bmvi .de/SharedDocs/EN/publications/report-ethics-commission-automated-and -connected-driving.pdf?__blob=publicationFile.

23. The names selected for the clusters weren't entirely fitting—the authors used an algorithm that grouped the preferences of participants from one country with those that were relatively similar from another. This method contained a large number of modeling decisions in its own right, so the "Western" group, in addition to many European countries and those of North America, also included Ban-

gladesh, Russia, and South Africa. Meanwhile, France and Hungary were included in the "Southern" section, alongside many countries in South America.

24. Awad et al., "Moral Machine Experiment," 61.

25. You can find general examples of such perverse incentives, where the actual behavior "perverts" whatever the ideal goal is, on Wikipedia: https://en.wikipedia .org/wiki/Perverse_incentive.

26. David Collingridge, *The Social Control of Technology* (New York: St. Martin's Press, 1982).

INDEX